Elements of Pattern Theory

Johns Hopkins Studies in the Mathematical Sciences

in association with

The Department of Mathematical Sciences,
The Johns Hopkins University

ELEMENTS OF PATTERN THEORY

ULF GRENANDER

Johns Hopkins University Press
Baltimore and London

© 1996 The Johns Hopkins University Press
All rights reserved. Published 1996
Printed in the United States of America on acid-free paper
05 04 03 02 01 00 99 98 97 96   5 4 3 2 1

The Johns Hopkins University Press
2715 North Charles Street, Baltimore, Maryland 21218-4319
The Johns Hopkins Press Ltd., London

Library of Congress Cataloging-in-Publication Data will be found
at the end of this book.
A catalog record for this book is available from the British Library.

ISBN 0-8018-5187-4
ISBN 0-8018-5188-2 (pbk.)

CONTENTS

Plates 1–11 follow page 130
Preface ix
Introduction xi

PART I. A Catalogue of Patterns

Chapter 1. The Search for Order — 3
  1.1 Knowledge and information — 4
  1.2 Knowledge representations — 5

Chapter 2. Open Patterns — 8
  2.1 Ornaments — 8
  2.2 Language patterns — 13
  2.3 A motion pattern — 21
  2.4 Yarns — 23
  2.5 Molecular chains — 23
  2.6 Time recordings — 26
  2.7 Plants and trees — 28
  2.8 Tracks — 29
  2.9 Behavior — 30
  2.10 Doctrines — 32
  2.11 Open patterns — 33

Chapter 3. Closed Patterns — 34
  3.1 More ornaments — 34
  3.2 Weaving — 38
  3.3 Textures — 44
  3.4 Shapes — 48
  3.5 Inside structure — 56
  3.6 Connections — 59
  3.7 Internal patterns — 66
  3.8 Multiple object patterns — 68
  3.9 Pattern interference — 70
  3.10 Pattern of speculation — 71
  3.11 Speculation about patterns — 76

PART II. Theory of Patterns

Chapter 4.		Analyzing Patterns	81
	4.1	Generators	81
	4.2	Configurations	83
	4.3	Images	91
	4.4	Patterns	93
	4.5	Probabilities	94
	4.6	Deformations	97
	4.7	Tasks of pattern theory	98
Chapter 5.		Analysis of Open Patterns	101
	5.1	Character strings	101
	5.2	Operator strings	104
	5.3	Finite state languages	106
	5.4	Context free languages	107
	5.5	Curve images	108
Chapter 6.		Analysis of Closed Patterns	112
	6.1	The Ising model	112
	6.2	Boundary patterns	119
	6.3	Deformable templates	122
	6.4	Restoring character strings	127
	6.5	Uncompromising logic	129
	6.6	Inference for deformable templates	130
	6.7	Abnormality detection	133
	6.8	Multiple objects	140

PART III. Computing Patterns

Chapter 7.		Computer Experiments with Patterns	145
	7.1	Programming language	146
	7.2	Hardware	147
	7.3	Programming strategy	147
	7.4	Some utilities	148

Chapter 8.		Computing Open Patterns	154
	8.1	Character strings	154
	8.2	FS and CF patterns	155
	8.3	Computing regime patterns	165
	8.4	Computing curve images	167
	8.5	Snake shapes	168
	8.6	Computing weaves	175
Chapter 9.		Computing Closed Patterns	178
	9.1	Computing Ising models	178
	9.2	Computing boundary models	180
	9.3	Computing deformable templates	182
	9.4	Star-shaped patterns	185
	9.5	Multimodal inference	190
	9.6	Computing character string restoration	194
	9.7	Growth patterns	199
Chapter 10.		More Pattern Experiments	201

Notes 211
References 217
Index 221

PREFACE

The search for patterns in nature and in the man-made world has generated a huge literature. Even if we include only quantitative studies – mathematical or computational – a complete list of references would have many thousands of items. Indeed, there has emerged a whole discipline, pattern recognition, that deals with the recognition of patterns by animals and machines. It is a discipline that has achieved many successes, both practical and theoretical, and it will no doubt continue to flourish.

This book has another orientation. We shall not investigate methods for recognizing patterns; at least this will not be our primary goal. Instead we shall try to formalize the very concept of a pattern in terms of a mathematical framework, a pattern theory.

To guide the reader towards the mathematical constructs that will be employed we first present a catalogue of patterns in Part I. Little mathematical knowledge is required of the reader of this part of the book.

In Part II we shall reason about these patterns, see how they have much in common even when they appear (superficially) quite different. Reaching this requires some knowledge of mathematics on the junior/senior undergraduate level.

In Part III we shall apply the pattern theoretic ideas to the construction of algorithms and computer programs that are intended to handle and analyze patterns. The reader with some computer experience should be able to carry out these experiments, at least with the help of a teacher. We shall start with simple experiments but lead up to some of considerable sophistication.

The work on developing pattern theory has been supported for several years by ARO, ONR, NSF and, more recently, by ARPA. The book is based on lectures given at Johns Hopkins University in 1991. I am grateful to Alan Karr for inviting me to give these lectures and to the Johns Hopkins University Press, represented by Richard O'Grady and Robert Harington. The preparation of the manuscript was facilitated by the efficient and competent help of Ezoura Fonseca. Some of the material was developed by my students as indicated in the text. To all the above I express my thanks.

INTRODUCTION

The subject of this book is order, patterns, regularity – concepts that imply that the world we live in has structure making it possible for us to understand it, at least to some extent. Without presupposing such a structure we would have no hope of comprehending the phenomena that we observe and the logical relation between them.

To make this concrete we shall present a *catalogue of patterns* in Part I. One extreme is a completely regular pattern – for example, a crystal – which can be explained through simple rules of generation. Another extreme is complete disorder in terms of pure randomness. We shall concentrate our attention on intermediate situations, in which the phenomena can be analyzed partially through notions of *typical structure* that may be obscured by a *high degree of variability*, as, for example, in many instances of biomedical images. Such patterns not only appear complex – *they are complex* – and therefore they differ in essence from, for example, fractals and various patterns in chaos theory that seem complicated although they may have been generated by comparatively simple rules.

Pattern theory is a way to approach patterns through a mathematical formalism, a way of *reasoning about patterns*. It will be presented in Part II, using analytical tools, and in Part III employing computational methods.

To avoid misunderstanding it should be pointed out that the book is not about pattern recognition, a discipline related to but different from pattern theory. The distinction is, with some exaggeration, that in pattern recognition the emphasis is on the construction of algorithm for *recognition* of patterns, while in pattern theory we concentrate on *the structure of patterns themselves*.

The book is intended for readers with a general interest in the fascinating world of patterns and who want a guide for exploring that world. Part I introduces such a reader to the basic concepts and ideas of pattern theory and is as non-technical as possible, assuming only some familiarity with elements of mathematics. Part II is more mathematical and assumes that the reader knows more about probability and statistics. Together with Part III it is intended for a reader who is interested in actually applying the theory to some particular application. Its more technical nature makes for harder reading, but an attempt has been made to avoid technicalities of interest only to the specialist.

The book is not intended as a detailed manual for applying pattern theory. Readers looking for that and who are prepared to deal with the necessary technicalities should consult sources listed in the references.

Computer experiments with patterns, as described in Part III, form a necessary supplement to the analytical treatment. The reader is encouraged to experiment on his or her own, and the computer programs offered in the text (and on a diskette that can be obtained from the author) can serve as a starting point. They are not particularly well written; little effort has been made to speed them up and none to structure the code systematically.

The book can be read in two ways. Either begin with Chapter 1, then go to Chapter 2 and so on:

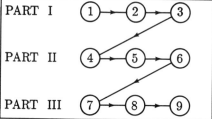

Or, start with the introductory and general Chapter 1, continue with the introductory mathematical Chapter 4 and computational Chapter 7. After these preliminaries go to the more detailed Chapter 3, etc. Like this:

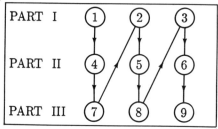

The author has taught many courses on pattern theory on the undergraduate level. It has been obvious while doing this that the computer experiments are indispensable for the students' understanding of the somewhat abstract concepts that appear. They can also be fun. The text has been developed for a senior seminar but could also be used as the basis for a graduate level course on the introductory level.

Some of the principles of pattern theory were announced already in the late 1960s but applications were not made until the early 1980s. One reason for this was that the theory was still quite incomplete, another that the computing power was not sufficient.

The ongoing revolution in information technology will continue to influence the development of the theory. It is not just a question of increasing computational speed, but advances in sensor performance (e.g. MRI, PET, laser radar) and information carriers (e.g. fiber optics, CD-ROM, networks) as well as in computer architecture (e.g. parallel machines, analog VLSI).

What earlier seemed to be unsurmountable computational obstacles can now often be overcome with reasonable cost and time. Workers in pattern theory feel a technological pressure, almost palpable, to exploit the new means by expanding and adapting the theory to increasingly complex patterns. It is a joy to be part of this intellectual adventure.

PART I

A CATALOGUE OF PATTERNS

Chapter 1

The Search for Order

The world we live in is not completely disordered, although it may sometimes seem so. Instead, it is governed by rules, some of which are exact, like the ones controlling the changes of the seasons and the eclipses of the sun, while others involve randomness, like Mendelian genetics or the behavior of animal collectives. A fundamental belief underlying all scientific endeavors is that the world can be explained and to some extent predicted in terms of such rules. Or, as Einstein once pointed out, the most remarkable fact about the universe is that it can be understood. Even prescientific thinking relies on such explanations. Thunder is caused by Thor throwing his hammer across the skies; the cows have stopped producing milk because the neighbor's wife has put a curse on them. Magic and mythology can be seen as early attempts to explain the world.

The idea that phenomena, even the most complex ones, should be understandable is not always accepted. It becomes especially sensitive and even controversial when applied to human behavior: we resist instinctively the idea that we can be analyzed and understood in essentially the same way, albeit on a more complex level, as when we take a mechanical device apart to see how it functions.

There is a tendency to ascribe intriguing events and phenomena to random causes, not amenable to exact analysis, at least not in mathematically precise form. On the other hand, we can sometimes see obvious patterns, typical behavior, more or less constrained by rules that can be clearly articulated.

We illustrate one of these opposing themes in Plate 1, Sturt's "The Battail of Nasbie" (courtesy of Brown University, Anne S. K. Brown Military Collection). Note the perfect order; the troops are arranged according to the rules of war in the seventeenth century. Carefully arranged marching columns, charging cavalry: a war scene of almost aseptic quality.

Contrast this with the painting in Plate 2 by Vassilii Vassilivich Vereshchagin: "The Battle of Borodino." He was famous for illustrating Tolstoy's *War and Peace*. It shows a terrifying picture of death and confusion, with corpses piled on top of each other. Tolstoy's view of history, including battles like this, is that of a succession of accidents with no apparent order; haphazard acts of individuals, here soldiers, make up the battle scene or more generally: randomness rather than order emerges as the dominating principle.

But the opposition of *order vs. disorder* is not a clear-cut one. No one, for example, expects a 48-hour weather prediction to be perfect; nevertheless, we put some trust in it and with good reason. Given the scattered information that the meteorological forecaster has available, and the sensitivity to changes in initial conditions, weather will be statistical in nature in addition to the deterministic laws of fluid mechanics. In spite of the random element we speak of weather patterns: prevailing westerly winds, a hot spell, rainy days. In everyday language we do not worry too much about the existence of these patterns or what we mean by the word "pattern." But what *do* we mean?

Perhaps we mean some basic structure behind the phenomenon we are watching, some more or less visible regularity, some underlying principles that tell us how the phenomenon can be reduced to simpler terms, primitives. In other words: a "typical structure" generated from rules to produce a regular appearance or behavior.

Perhaps we also allow variations on this typical structure, making it look at least superficially different at different occurrences. "Different," in opposition to "same," could here be something trivial – change in location, scale and orientation in a picture or in time – man does not enter the same river twice, to quote Heraclitus. Or it could be something more fundamental – in one CAT-scan (Computerized Axial Tomography or Technique, discussed in section 3.6) of a brain the pituitary gland appears normal while in another it has atrophied. This leads us to describe and formalize patterns by rules for typical structure but also rules for variability: these two opposing themes have to be combined.

Some patterns are rigid and easy to understand with little analytical effort. Such are many man-made structures like textiles or programming languages. They can be understood by the weaver or software engineer who created them and can tell others how they were made. We shall pay some attention to such rigid patterns, but only as a preparation for more challenging ones from nature, for example biology and medicine, where variability is all important.

The analysis will be made using the mathematical constructs in Part II, *regular structures*, and will formalize subject matter knowledge from the field we happen to be concerned with. But first let us reason in more intuitive terms.

1.1. Knowledge and information. What sort of knowledge do we have in mind? The phone book for Providence, Rhode Island, contains a mass of information, but its content scarcely deserves to be called knowledge. The Yellow Pages are structured into classes, such as Book Stores, Used Automobiles, Physicians. This is taxonomy: a list of "things" divided into sets in some useful way. It is like botanical taxonomy, for example the Sexual System introduced by Carolus Linnaeus for classifying plants based on pistils and stamens first, other features second. A taxonomy can be useful without being profound, but it represents only limited insight.

It is only when information has been structured in a logical and connected way that the term knowledge deserves to be applied. An almanac listing the times of sunrise and sunset or the position of the planets is useful. Kepler's laws or the rules of Newtonian mechanics, however, make us *understand* the data in the almanac in a more satisfying way; it is deeper since it reduces a wealth of information to basic principles.

Scientists sometimes speak of reductionism, often in a pejorative sense, the philosophical attitude that tries to explain complex systems in terms of concepts and ideas that are so basic, so primitive, that they appear as indivisible, not needing any further motivation. To explain biology in general by reference to the Schrödinger equation would be a drastic example of a reductionism that seems (presently) hopeless and is viewed as too ambitious by many biologists. On the other hand, the desire to reduce genetics to molecular biology would be an example of reductionism that seems to be succeeding at present.

When we encounter a body of knowledge our goal will be to reduce it to primitive concepts and to formalize the doctrine that it embodies in precise categories. Knowledge will be thought of as descriptions, patterns, often pictorial, sometimes linguistic, but also in many other appearances. The main difficulty will then be to introduce, invent if you will, meaningful primitives and rules for manipulating them into formal structures that at least formalize the body of knowledge we are studying and do so in precise language. This is a tall order!

1.2. Knowledge representations. Our program is therefore to create a theory of patterns that can be used systematically to codify knowledge at least in many situations. This is similar to *knowledge engineering* in artificial intelligence (AI), where computer scientists try to create representations (for example of a medical doctrine) that are possible to exploit by implementing them on the computer. Our approach differs in that it will be based on an algebraic/probabilistic foundation that has been developed for exactly this purpose. The algebraic component will express rules of regularity, the probabilistic one the variability, thus combining the two opposing themes mentioned above.

In the history of mathematical modeling of the world the outstanding success story is in physics, beginning with mechanics. Take for example Galileo's work on the laws of falling bodies. With audacious disregard of "reality" he postulated that the fall took place in a room without atmosphere, no air. He speculated in the famous discussion between Salvatius and Sagredus (note 1.1) on how bodies *ought* to fall and supported this, at least to some extent, by direct observations at the leaning tower of Pisa. The law he formulated was that the vertical distance z traveled by the falling mass in time t could be written, in modern mathematical notation, as

$$z = \frac{1}{2}gt^2$$

where the constant g is the earth acceleration constant. This law, which at the time seemed to contradict common sense (common sense dictates that heavy bodies should fall faster than light ones), was arrived at by what is nowadays known as *Galilean simplification*. Instead of studying a problem with all its complications and qualifications he abstracted from all but a few of the characteristics of the problem. To make that abstraction so that a problem becomes solvable without losing too much of its substance is an art for which few general principles can be given.

As mechanics evolved after Galileo it led eventually to Newtonian mechanics, one of the most magnificent achievements of the human mind. In Newton's clockwork universe, cold and beautiful, the planets move around the sun governed by three simple principles; all mechanical movements can, at least in theory, be derived from them.

Laplace, eminent mathematician and a less successful minister of finance to Napoleon (note 1.2), suggested that if there were an all-seeing being, the so-called Laplace's demon, who could observe the positions and velocities of all material particles in the universe at one time, then it would be possible to predict the future development of the universe without error and for all times in the future (as well as in the past). Today one would not take this very seriously, but the thought illustrates the belief in the supreme power of mathematics as a tool for understanding the universe. Outside of physics the exploits of applying mathematics have been less spectacular but with many successes. Let us just mention the insurance business, whose mathematical basis is probability theory, or numerical weather prediction that rests on the differential equations of fluids and perhaps thermodynamics. In the life sciences the last several decades have witnessed a multitude of attempts to describe (for example) biological development and evolution in precise, quantitative terms. Many biologists react with scepticism, sometimes justified, to such attempts.

Biology is not physics. The objects of study, organisms on different levels, are thought to be the result of a long, more or less haphazard sequence of events that influenced evolution. If this is so there is no reason to believe that the result, for example the DNA sequences, is simple in any real sense; Galilean simplification will not work.

One could try to avoid the full impact of this pessimistic proposition by referring to fractals and theory of chaos (note 1.3). Fractals are endlessly repeating figures and form sets in some space, for example the plane; they can be of great beauty and may appear extremely complex. They can be generated in many ways, one of which is by repeated application of transformations of the plane. What is perhaps most remarkable is that some of them show an uncanny resemblance to some living organism, for example ferns or trees. And this in spite of the fact that the transformations are quite simple. They *appear* complicated but are simply generated.

Our aim is different. *We will try to represent patterns that appear complex and are complex.* For those the representations will have to be complicated, perhaps incorporating megabytes of constants, not just a few parameters as we are used to in most mathematical models in physics. If this can be done and if the representations can be made mathematically stringent and computationally feasible we will have a powerful tool for pattern research in biomedical imaging, to mention just one of the most exciting possibilities. The development of pattern theory is a major intellectual challenge and promises huge practical payoffs.

We shall start, however, with much simpler patterns, see how they can be represented and processed, and then proceed to patterns of gradually increasing complexity.

Chapter 2

Open Patterns

Let us now begin to search systematically for regularities that can be used for the representation of knowledge in the form of patterns. We will be confronted by formidable difficulties, both conceptual and mathematical, so let us start with patterns whose internal structure is as simple as possible. By "simple" we mean patterns whose logical architecture – its connectivity – does not involve recurrencies (loops) but is straightforward. We shall postpone giving a definition of this until section 2.11 and more formally in Part II.

2.1. Ornaments. An obvious and rich source of pictorial patterns can be found in the visual arts. This is true for traditional paintings as well as for non-representative ones. For our purpose the ornamental arts are best suited to illustrate the regularities that we are searching for. Consider, for example, the ornament in Figure 1. It is from the magnificent collection of ornaments by Owen (1972). This is one of the simplest ornaments in Owen's volume and its structure is obvious: the shape ⌐⌐ is repeated several times, each time shifted to the right. The shapes are joined together so that they form a continuous curve.

FIGURE 1. GREEK ORNAMENT

A seemingly more complicated example is shown in Figure 2. What regularities seem to be present, how are the images generated? Looking closely at the graphs one notices that some elements appear again and again.

Chapter 2: Open Patterns

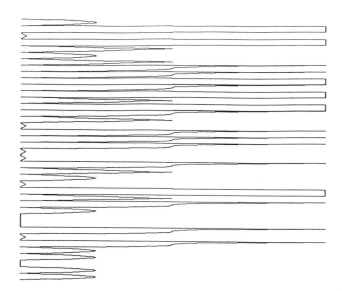

FIGURE 2. SYNTH1

Collecting all such elements we will find the six represented schematically in Figure 3 and we shall refer to them as a,b,c,d,e,f, coded into numbers 0,1,2,3,4,5.

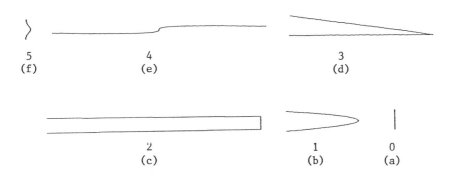

FIGURE 3. ORNAMENT ELEMENTS

10   *Chapter 2: Open Patterns*

Coding the images in Figure 2, reading it from below, we get the sequence 1 1 0 1 1 4 5 4 0 0 1 4 3 2 5 1 3 4 5 5 4 4 4 5 3 4 2 3 2 4 2 4 4 3 1 3 2 5 2 1, with no apparent order in the arrangements; it seems to be randomly generated. And it is.

It is quite different with the image in Figure 4.

FIGURE 4. SYNTH2

This is coded into 0 1 2 3 4 5 0 1 2 3 0 1 2 3 4 5 0 1 2 3 4 5 0 1 2 3 4 5 0 1 2 3 4 5 0 1 2 3 4 5. It is periodic, with the period abcdef = 012345 repeated. Simple enough!

Now look at Figure 5, coded into 0 2 5 3 2 3 4 4 1 3 4 4 1 3 3 3 0 2 5 3 2 3 4 4 1 3 3 3 0 2 5 3 2 3 4 4 1 3 3 3. They are also periodic but with the period bdddacfdcdee = 133302532344. We designed this example so that it would be easy to distinguish the elements that make up the pattern. Once this has been made we can code the images into strings over some alphabet, say the set of numerals or numbers, making it easier to discern regularities. In general, however, *it will be harder to find (or create) the elements*: this is a first and often challenging problem in the study of patterns.

Let us look more carefully now on the families of strings in the last two examples, and let the strings be very long and written as $x_1, x_2, x_3, \ldots x_L$, L big, so that any x_i is one of $a, b, c, \ldots$. What are economical representations of such strings? We can certainly use

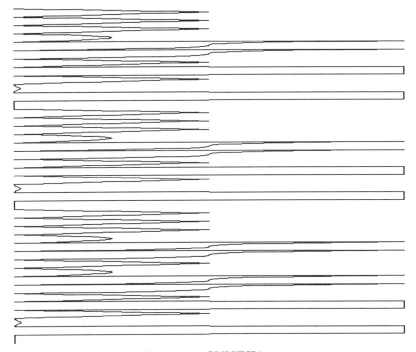

FIGURE 5. SYNTH3

the string itself which takes L spaces in memory, each space meaning one character, one byte. This is a waste of space since the strings are periodic, say with a period p, where $p = 6$ in the first example and $= 12$ in the second. It would be more economical to store in memory the period, say, $bdddacfdedee$, and the initial element x_1.

This seems to work in the example. However, if we want $x_1 = d$ we do not know if x_2 should be d, c, e, or a, so that the representation is not adequate. We can remedy this by using a counter, call it k using the correspondence

$$\begin{array}{cccccccccccc} b & d & d & d & a & c & f & d & c & d & e & e \\ \downarrow & \downarrow & \downarrow & \downarrow & \downarrow & \downarrow & \downarrow & \downarrow & \downarrow & \downarrow & \downarrow & \downarrow \\ 0 & 1 & 2 & 3 & 4 & 5 & 6 & 7 & 8 & 9 & 10 & 11 \end{array}$$

and store the whole period and the initial $k = k_1$, value where the first x_1 corresponds to k_1. This representation is unique.

If we write as $x = f(k)$, so that $f(0) = b, f(1) = d, \ldots$, then the x-string can be obtained as the function f applied to the k-string. This means that the k-variable plays the role of a *counter* and the k-string is a *coordinate* axis, although of somewhat unfamiliar type. Note that the above correspondence is not one-to-one: a k-value determines the x-value, but

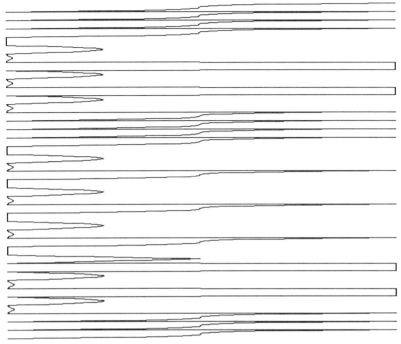

FIGURE 6. SYNTH4

$x = d$ corresponds to the five k-values $1, 2, 3, 7, 9$. This innocuous observation will come back to haunt us.

But periodic patterns are really too simple to engage our imagination. Look at the image in Figure 6: it is not periodic; what is its regularity? This is harder but it will be seen that the string eeefbcfbcdaefbaefbaefbaeeeefbcfbcfbaeeee satisfies the rules

$$\begin{cases} a \text{ is followed by } e \\ b \text{ is followed by } a \text{ or } c \\ c \text{ is followed by } d \text{ or } f \\ d \text{ is followed by } a, e \text{ is followed by } e \text{ or } f, f \text{ is followed by } b \end{cases}$$

This allows for more variability in the family of strings but the constraints are still fairly simple, once discovered, since x_{i+1} is only restricted by x_i: *local constraints*, only the left neighbor has direct influence on x_{i+1}. Actually, this dependence propagates, so that for a trimer (x_i, x_{i+1}, x_{i+2}) the choice of x_{i+1} is restricted both by its left neighbor x_i and the right one x_{i+2}. For example, if $x_i = b$ and $x_{i+2} = e$ only $x_i = a$ is allowed; see above list of constraints.

We shall return to ornaments, but in two dimensions, in Chapter 3.

2.2. Language patterns.

Ornaments are man-made and more or less conscious products of an artistic mind, so that it ought to be possible to understand/represent them; in many cases this is quite easy. It is different with languages; they have grown gradually and not under full human control. This is true even when such control is attempted as when the Académie Française has tried, in vain, to isolate French from foreign influences. Natural language patterns are among the most complicated and hardest to understand.

The attempt to understand natural (as contrasted to formal) language in a scientific manner is at least twenty-five centuries old, going back to Panini's remarkable grammar of Sanskrit (see note 2.1), the rigid and idealizing Greek-Latin grammars, the German philological school from the nineteenth century, and "modern" syntax beginning with de Saussure's celebrated "Cours de linguistique general."

Behind all these attempts is the tacit assumption that language structure can be understood, that it is a complex but rational system whose patterns can be analyzed in a logical way in terms of primitives: word classes, parts of speech, inflection rules, laws of agreement (in number, gender, ...), or parsing schemes. In spite of these glorious efforts language still remains something of a mystery whose structure is only partially understood (see note 2.2). Our goal here is a more modest one, only to look at some linguistic ideas and use them to help build regular structures applicable in other contexts.

2.2.1. It is best to start with an extremely simple language fragment: the sort of sentence that would occur in a child's first reader. Consider sentences like

$$\left\{ \begin{array}{l} \text{Mary and John play} \\ \text{John sees a big house} \\ \text{father drives his car to work} \\ \text{George has a red tricycle} \\ \text{they are happy} \\ \text{the big boy chases the little girl} \\ \ldots \end{array} \right.$$

Can we find any regularities in them? We know from grammar learned in school that words come in different forms and belong to different word classes. It seems obvious (obvious but wrong!) that the elements from which language patterns are built are the words, so let us think about words as primitives, atoms of language. Much better primitives will be discussed later. Word morphology divides words into classes, for example

$$\left\{\begin{array}{l}\text{personal pronouns PP: they, he, }\ldots\\ \text{personal names PN: John, Mary, }\ldots\\ \text{conjunctions C: and, or, }\ldots\\ \text{adjectives Adj: red, big, }\ldots\\ \text{nouns N: house, car, tricycle, }\ldots\\ \text{transitive verbs }Vtr\text{ : sees, drives,}\ldots\\ \text{intransitive verbs }Vitr\text{ : sit, }\ldots\\ \text{article Art: the, a, an, }\ldots\\ \ldots\end{array}\right.$$

or preferably into more delimited subclasses. In a word class, for example nouns, different forms may appear: singular, plural, possessive; in some languages also gender. For verbs: first person, second person, singular, plural, present, future, and many other forms.

In addition to this morphological aspect words can also be viewed from a frequency point of view: some words are used more often than others. If we enumerate the words used as $w_1, w_2, w_3, \ldots$ we may associate probabilities $p_1, p_2, p_3, \ldots$ to the respective words. In a large corpus of text, say with N words, we then expect approximately Np_1 occurrences of the word w_1, Np_2 of the word w_2, and so on.

In a famous experiment Shannon (1949) used a set of probabilities $\{p_1, p_2, p_3, \ldots\}$ to simulate or, to use a more appropriate term, *synthesize* a fragment of text:

Representing and speedily is a good apt or come can different natural here the a in came the to of to expert gray come to furnishes the line message had be these.

It does not look much like English.

The reason why this fragment appears chaotic is that successive words have been chosen at random independently of the earlier ones. To remedy this Shannon collected conditional words probabilities

$$p_{ij} = P(\text{word } w_i \text{ occurs if the previous word was } w_j),$$

where P(event) denotes the probability of the event. Using all the $\{p_{ij}\}$ he synthesized this *Markov chain* (note 2.3) and got

The head and in frontal attack on an English writer that the character of this point is therefore another method for the letters that the time of who every told the problem for an expected.

This looks more like English. To get even better similitude one could perhaps use conditional probabilities expressing two-word dependencies:

$$p_{ijk} = P(\text{word } w_i \text{ occurs if the two previous ones were } w_j \text{ and } w_k).$$

The trouble with this is that the table $\{p_{ijk}\}$ is going to be enormous, and if we include dependencies on even earlier words, or words far apart in the sentence (in German the verb is sometimes at the end referring to a subject much earlier) the memory size needed is going to be prohibitive.

A deeper criticism of this attempt to represent language patterns is that it is too mechanical, it does not penetrate the substance of the problem in the way that grammars (see below) at least attempt to do. Nevertheless these probabilistic language representations have their practical use as we shall see later.

It should be pointed out that the probabilities will depend upon what corpus of text we are dealing with. This fragment is in a learned *style*:

Less of a commonplace is, I think, the ways in which the inherent tension between these two legacies is reflected in the turbulent technological developments, the crisis of spiritual values, and the threats even to man's physical survival which have become characteristic of the twentieth century in its final decades.

This fragment is more colloquial:

"They's no use kiddin' ourself and more," said Tommy Haley. "He might get down to thirty-seven in a pinch, but if he done below that a mouse could stop him. He's a welter; that's what he is and he knows it as well as I do. He's growed like a weed in the last six mont's. I told him, I says, 'If you don't quit growin' they won't be nobody for you to box, only Willard and them.' He says, 'Well, I wouldn't run away from Willard if I weighed twenty pounds more.' "

We can expect their probabilistic representation to be quite different.

A less extreme example with less clear-cut differences is in the Federalist Papers, a collection of essays debating the Constitution of the United States to be adopted. The papers are supposed to be written by Alexander Hamilton, James Madison, and John Jay, but the attribution for some of them is in doubt. In Mosteller-Wallace (1964) an attempt is made to achieve objective determination of authorship, representing literary styles by word probabilities. This can be viewed as literary pattern recognition.

Let us return to the first grade reader. Some regularities are apparent. For example, the first sentence above can be changed into

$$\begin{cases} \text{Peter and John play} \\ \text{Peter and Cindy play} \\ \ldots \end{cases}$$

More abstractly we can represent these sentences as

$$PP \text{ and } PP \ V_{itr}$$

standing for all sentences of the form

$$\begin{cases} x \text{ and } y \ z \\ x \in PP, \ y \in PP, \ z \in V_{itr} \end{cases}$$

The sentence we started with, call it the *template*, has been changed into many others by a family of substitutions where a word is replaced by an arbitrary word from the same word class. The sentences in the resulting family can be said to be grammatically similar.

But this regularity is restricted; we need much more. Consider the sentence

Mary's little doll is cute

in which the word "doll" is qualified by the two adjectives "little" and "cute." What linguistic device (or convention) is used to represent these two references to Mary? "Little" immediately precedes "doll," which in some languages serves as such a device. In some languages the adjective could come after the noun but the idea is the same: word order can be exploited for references.

But how about the reference "cute" → "doll"? In languages more inflected than English, for example Latin, it could be made more explicit using endings as relation indicators: praeterea censeo Carthaginem esse delendam where "delendam" refers to "Carthaginem." If we want to express the references in

Mary's little doll is cute

and were free to invent linguistic devices one could try to introduce a coordinate system in the sentence, coordinate system understood in a very general sense a bit like the one with k-values that we used for ornaments. Just as when we go along the ornament we successively unravel the coordinates. We can use rules to do this by a *finite state, FS, grammar*.

Let us explain what this means by a simple example; in Part II we shall make this more precise and formalized. We need three concepts, the first of which is a list of words, the *terminal vocabulary* denoted V_T. Let us choose V_T as the union of the word classes

$$\begin{cases} \text{PN} & = \{\text{John, Peter, Mary}\} \\ \text{PPS} & = \{\text{he,she}\} \text{ (singular)} \\ \text{PPP} & = \{\text{they}\} \text{ (plural)} \\ \text{VITRS} & = \{\text{plays,sings,runs}\} \text{ (singular)} \\ \text{VITRP} & = \{\text{play,sing,run}\} \text{ (plural)} \\ \text{CONJ1} & = \{\text{and}\} \\ \text{CONJ2} & = \{\text{but,while}\} \\ \text{PUNCT} & = \{\cdot\} \text{ period} \end{cases}$$

We also need coordinates, here called *states*, forming the *non-terminal vocabulary* denoted V_N. We shall choose
$$V_N = \{1,2,3,4,5,6,7\}$$
where 7, denoted by F in a general situation, will be called the *final state*. The *initial state* will be 1.

Finally we need a set of *rewriting rules* that tell us what to "write" as we go from one state to another. Let us choose the rules

Table 1	
$1 \xrightarrow[PN]{} 2$	$5 \xrightarrow[CONJ2]{} 1$
$1 \xrightarrow[PPP]{} 2$	$1 \xrightarrow[PN]{} 6$
$2 \xrightarrow[CONJ1]{} 3$	$1 \xrightarrow[PPS]{} 6$
$3 \xrightarrow[PN]{} 4$	$6 \xrightarrow[VITRS]{} 5$
$4 \xrightarrow[VITRP]{} 5$	
$5 \xrightarrow[PUNCT]{} 6$	

This is intended as follows. We always start generating a sentence in State 1. We can go from State 1 to State 2 writing "John" or "Peter" or "Mary", that is one of the words in the class PN. But we can also go from State 1 to State 2 writing "they," the only word in the class PPP, and so on. When we get to the final state, in this case $F = 7$, we stop, the sentence is complete.

It will be convenient to visualize Table 1 as the *wiring diagram of a finite state machine* as in Figure 7.

Let us run the finite state machine. Start at 1 and run through 2,3,4,5,7:

$$1 \xrightarrow[Mary]{} 2 \xrightarrow[and]{} 3 \xrightarrow[John]{} 4 \xrightarrow[play]{} 5 \xrightarrow[.]{} 7$$

Or, run through 1,6,5,1,6,5,7:

$$1 \xrightarrow[he]{} 6 \xrightarrow[plays]{} 5 \xrightarrow[but]{} 1 \xrightarrow[Mary]{} 6 \xrightarrow[sings]{} 5 \xrightarrow[.]{} 7$$

These sentences look right but we can also get

$$1 \xrightarrow[John]{} 2 \xrightarrow[and]{} 3 \xrightarrow[John]{} 4 \xrightarrow[run]{} 5 \xrightarrow[.]{} 7$$

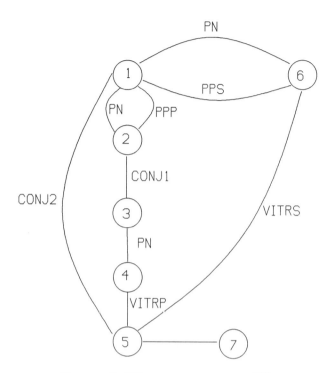

FIGURE 7. WIRING DIAGRAM OF FS MACHINE

which may be grammatically correct but sounds funny. A more substantial criticism is that the FS languages generated by finite state grammars are too restricted. One can extend them by adding rules to be more realistic, but the number of rules required to get reasonable approximations to natural language will be prohibitively large. The elements used to generate the FS languages are the rewriting rules. Our earlier statement that the words should be chosen as elements may have been obvious but is inadequate. In an FS grammar we are free to choose successive rewriting rules at will as long as a rule $i \xrightarrow{x} j$ is followed by one of the form $j \xrightarrow{y} k$; otherwise no restrictions apply. Note that this single restriction is a sort of continuity constraint: we do not allow jumps in $j \to j'$, only $j \to j$. The resulting word string, however, can have complicated dependencies.

Another comment, seemingly obvious, that will later be seen to be fundamental is that the sentences are what can be observed: heard or read. The generating string of states can (perhaps) be derived, computed, but is not immediately observable.

2.2.2. To illustrate the limited generative power of FS grammars let us look at the

sentence
$$\underbrace{\text{the little boy}}_{NP}\ \underbrace{\text{who saw a dog}}_{RC}\ \underbrace{\text{was afraid}}_{VP}$$

or expressions from programming languages
$$\begin{cases} x+y \\ (3*u-v) \div 5 \\ x*v+y*(3*x-v) \\ \ldots \end{cases}$$

Such sentences are awkward to analyze by FS grammars. It is more natural to use *context free*, CF, grammars. To analyze such expressions we introduce a terminal vocabulary
$$V_t = \{x, y, z, \ldots\} \cup \text{ a set of numbers } \cup \{(,)\}$$
where the set of numbers could be, for example, a set of integers and the last set consists of left and right parentheses. Let the non-terminal vocabulary be simply $V_n = \{S\}$ and define rewriting rules
$$\begin{cases} S \to S+S \\ S \to S-S \\ S \to S*S \\ S \to S \div S \\ S \to (S) \end{cases}$$

When we derive a sentence we no longer get a linear chain as for FS languages: branching creates a tree-like diagram as in

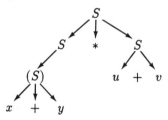

generating the arithmetic expression $(x+y)*u+v$. Similarly we generate a sentence by the tree-like diagram

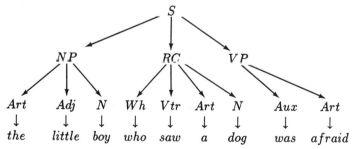

where we have used rules
$$\begin{cases} S & \to NP \ RC \ VP \\ NP & \to Art \ Adj \ N \\ RC & \to Wh \ Vtr \ Art \ N \\ VP & \to Aux \ Art \\ Art & \to the \\ Art & \to a \\ \ldots \end{cases}$$

where we have indicated the noun phrase "the little boy," the reflexive clause "who saw the dog," and the verb phrase "was afraid." We could also generate this sentence by an FS grammar, perhaps starting with the templates

$$\begin{cases} NP & = Art \ Adj \ N \\ RC & = Wh \ V_{tr} \ Art \ N \\ VP & = Aux \ Art \ N \end{cases}$$

where Wh stands for a word class {who,that,...} and Aux = {is,are,was,were...}. In the finite state machine diagram we then need one path that generates successively NP, RC, VP to do the job. This was to generate one type of linguistic behavior (reflexive clause). When we try to deal with another we can add more paths in the machine diagram – the model represented by the diagram will grow more and more until it becomes remarkably ugly and involved.

We should strive for parsimony in the representations of patterns, get deeper into the language structure. Concepts like NP, RC, VP seem natural (at least our schooling has conditioned us to think so) and they should be elements in the grammar. But if we try rewriting rules of the form

$$S \to NP \ RC \ VP$$

we have rewritten one state S into three "states," not as in FS rules where a state i is rewritten into at most one state j.

This led grammarians to allow branching (from one state into more than one) in the rewriting rules. To be concrete, say that we want to generate the language of simple algebra. The language will also contain expressions like $(x+y) \div (x+y)$ that could be simplified, and meaningless ones, like $x \div 0$. Nevertheless it does represent the sort of expressions we wanted. A grammar of this type is called *context free*, CF and clearly is more powerful than FS: any FS language can be represented as CF, but not always the other way around. See also note 2.2.

2.2.3. For a given natural language, say, English, it would be naive to search for a single grammar that would represent all the sentences in the language, from geographic region to region, in time, and between genres. The variability becomes an even more unavoidable issue when it comes to spoken, colloquial language. For example, in an ordinary telephone conversation the sentences are left unfinished, words are repeated and all sorts of phrases are used that normally would not be accepted. To represent colloquial language one could try to add rules expressing such "mistakes," but, as has been mentioned above, this will lead to an unmanageably large set of rules.

An alternative would be to stay with a grammar that approximately represents the language fragment that we are dealing with, but allow mistakes in the sentence derivation with some probabilities. For example, if we use a CF grammar and have the rules

$$\begin{cases} NP \to Art\ N \\ NP \to ART\ Adj\ N \\ NP \to N \end{cases}$$

we may, perhaps, allow also the derivation

$$NP \to Art\ Adj\ N\ N$$

with a small probability in order to represent repetition of words, although this rule is not in the original grammar. We would also allow for incomplete derivations, and so on. If we do this the distinction of grammatically correct/incorrect becomes less sharp. Instead of this dichotomy we will have a quantitative differentiation where sentences are associated with probabilities that express their *degree* of grammaticality. We shall see for many types of patterns how such a quantification is forced upon us when we deal with highly variable patterns.

2.3. A motion pattern. It must have been noticed early in most cultures that the stars could be seen as fixed, perhaps attached to some invisible sphere, and rotating around an axis through the polar star. The motion of the sun and the moon could also have been described by circular motions with the earth in the center. Indeed, this fits in well with Plato's belief in circular motion as the ideal one, but the problem was to reconcile this belief with the seemingly irregular, back-and-forth motion of the other planets: "to save the appearances," to use a common phrase. Observing for example Jupiter, and plotting the observed successive position against a star chart, one finds that the planet occasionally reverses its motion: this is called retrogression.

To account for such disturbing anomalies the classical astronomers had to modify a purely Pythagorean universe with a few spheres inside each other to what was to become the model described by Ptolemy in his Almagest. The idea was to preserve the circular motion as the basic assumption but combine such motions into compound ones. A circle is made relative to another so that a point of the first one will move along an epicycle. In this way the resulting motion will sometimes appear as retrogression explaining the anomaly.

A Ptolemaic universe could look like the picture in Figure 8, where the earth is in the center, the moon and the sun rotate around it in slightly eccentric circles, and the five planets follow epicycles. This magnificently conceived system of the universe enabled the astronomers to make numerical predictions of considerable accuracy. As astronomical observations became more accurate the Ptolemaic model had to be refined to reconcile it with data, and this was done by adding more circles. A late version of the model had thirty-nine circles. As these models developed from the simple one in Figure 8 to more complex ones, the basic idea has been to combine certain given motions – circular motions – with one another. A uniform circular motion is determined by the plane in which it is carried out, its center in the plane, its radius and its angular velocity; it has seven parameters.

When a new circle is added to the system its center is positioned on the periphery of another one. The appearance of the resulting system is what earthbound observers can see as time goes on. Their view of the planetary system is related to but conceptually distinct from the system (model) itself, a distinction that will reappear many times below.

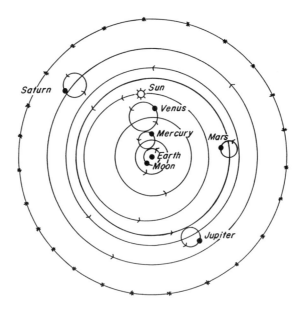

FIGURE 8. PTOLEMAIC SYSTEM

In the Ptolemaic system one can think of the circular motions as the primitives; these primitives are connected together with rules just described. The whole system is a combination of elementary concepts, a bit like the way a molecule is a combination of atoms following physical laws.

2.4. Yarns. Some of the most beautiful patterns occur in textiles. Most of them are two-dimensional and will be discussed in Chapter 3, but others, especially yarns, are essentially one-dimensional and we shall make some brief remarks about them here.

A yarn is made up of fibers that are combined during the drafting process which can be approximated as follows. A sliver is moving along its axis at uniform velocity V_1 and under one roller B (see Figure 9). Another roller, F, suddenly accelerates it to a higher velocity V_2. Just when the change occurs may vary in an irregular manner due to several factors, one of which is the interaction between adjacent fibers.

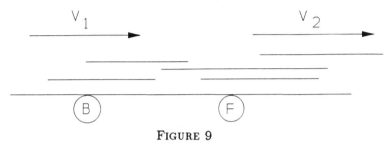

FIGURE 9

If we place a coordinate axis, labeled y, in the main direction of the yarn we can count the number of fibers $n(y)$ at location y. The function $n(\cdot)$ can vary regularly as in Figure 10(a) or in a more random manner as in (b). The behavior of the function will decide the appearance and strength of the yarn. Here we have neglected the three-dimensional aspects of a yarn; during the drafting process fibers will be made to rotate around each other.

The elements of the yarn (the fibers) obviously interact with each other, otherwise they would not hang together. The interfiber friction is too complicated to be described here; suffice it to say that these couplings are what makes a cohesive whole out of the collection of fibers.

2.5. Molecular chains. In section 3.2 we shall briefly consider some classical patterns of chemical configurations that describe molecules that can be quite complicated, as are the DNA (or RNA) chains that express genetic information.

DNA usually occurs in the form of two strings, the double helix, that complement each other. These strings are written over a small alphabet with four nucleotides:

$$\begin{cases} A \text{ for adenine} \\ C \text{ for cytosine} \\ G \text{ for guanine} \\ T \text{ thymine} \end{cases}$$

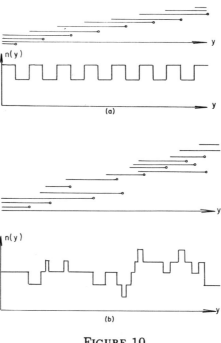

FIGURE 10

The Watson-Crick pairing says that in the two strings G bonds with C and A with T, so that the double string could look like

$$\cdots A \to G \to C \to T \to T \to T \to C \to T \to A \cdots$$
$$\cdots T \leftarrow C \leftarrow G \leftarrow A \leftarrow A \leftarrow A \leftarrow G \leftarrow A \leftarrow T \cdots$$

This double chain contains the base pairs $A - T, G - C, C - G$, and so on. The number of base pairs can be very large in some species, say, approximately 3.10^9 for human DNA strings. Since the two strings correspond in a one-to-one manner to each other we lose no information if we consider only one of them.

During evolution a string is copied but with occasional errors (mutations). A string s' descendant after generations from a string s'' can therefore be expected to be close (but distinct) in some as yet unspecified sense to s''. If the intervening number of generations is small one would expect s' to be quite close to s'' and vice versa. It is therefore reasonable to base the determination of genealogies for a set of strings on mutual differences. But how should distance be defined here? Much attention has been given to this and we shall describe only one approach.

Let us abstract a bit from the above and say that we have two strings

$$\begin{cases} a = a_1 a_2 \ldots a_n \\ b = b_1 b_2 \ldots b_m \end{cases}$$

over the same alphabet A of symbols; a's and b's from A. If the strings were of equal length we could use for example the *Hamming distance*

$$dist(a, b) = |\{a_i \neq b_i;\ i = 1, 2, \ldots n\}|$$

that is the number of pairs $a_i \neq b_i$ to measure how different a is from b. But we can have $n \neq m$ and what do we do then?

Let us think of successive changes $a = a^{(0)} \to a^{(1)} \to a^{(2)} \ldots \to a^{(r)} = b$ where for each time t the string $a^{(t)}$ is changed into some string $a^{(t+1)}$

$$a^{(t)} \to a^{(t+1)}$$

Let us think of a family of *simple moves* for elementary *deformations* in one time step. We could define the family $\mathcal{D}$ of such deformations to consist of

$$\begin{cases} \text{substitutions:} & \text{a letter } x \text{ is replaced by a letter } y \\ \text{deletions:} & \text{a letter } x \text{ is left out} \\ \text{insertions:} & \text{two consecutive letters } xy \text{ are replaced by the triplet } xzy \end{cases}$$

and perhaps others.

To each elementary deformation d we associate a positive number $e(d)$, the *effort*, and assume additivity, meaning that the total effort in the chain of deformations is defined as

$$\begin{cases} e(d_1) + e(d_2) + \ldots + e(d_r) & = e(D) \\ d_1 \text{ deformation used in} & a^{(0)} \to a^{(1)} \\ d_2 \text{ ---''---} & a^{(1)} \to a^{(2)} \\ \vdots & \\ d_r \text{ ---''---} & a^{(r-1)} \to a^{(r)} \\ D = (d_1, d_2, \ldots d_r) \end{cases}$$

Then define the distance between a and b as

$$\begin{cases} \min_D e(D) \\ D \text{ takes } a \text{ into } b \end{cases}$$

the minimum taken over all deformation D that change $a = a^{(0)}$ into $a^{(r)} = b$. This distance definition, the so called *Levenshtein distance*, is somewhat arbitrary but has some attractive features.

This implicitly introduces a *pattern dynamics* for strings if we assume that evolution (dynamics) follows the principle of minimum effort: from pattern a to pattern b the evolution tries to minimize effort. This has a reassuring similarity to the principle of least action in rational dynamics but it is no more than a striking analogy. It uses the philosophical idea of Occam's razor to select the simplest (smallest effort) of the evolutionary paths from pattern a to pattern b. This tries to give the simplest explanation of why the observed string a was deformed into another observed string b. A related question is how an initial string a is *likely* to evolve; in order to make this precise we must give a formal definition of "likely"; more about this in Chapters 5 and 8.

2.6. Time recordings. The patterns that we have examined so far are all one-dimensional in a sense that will be made mathematically precise in Part II. Among the most striking one-dimensional patterns are the electro-physiological recordings: EKG (electrocardiogram), EEG (electroencephalogram), and EMG (electromyogram).

In an EKG one records the skin potentials caused by the electrical activity of the heart. To do this one places electrodes at certain standard locations, the leads on the body, and measures the potential differences. The doctrine, apparently correct, is that the resulting patterns of the time recordings supply useful information about the patient's heart.

In Figure 11 we show some "normal" recordings for a particular placement of electrodes. A cursory inspection of the graph shows that the variations are, if not periodic, at least close to being periodic.

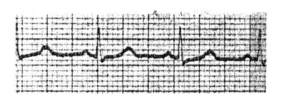

FIGURE 11. NORMAL EKG

Actually, a "period" looks somewhat like Figure 12 in which the subgraphs have generally accepted names: the P wave, the QRS complex, and so on. These subgraphs correspond in a qualitatively understood way to various electrical events (polarization, depolarization, etc.) in the heart.

In Figure 13 we show an abnormal EKG where no P-wave appears. To the trained cardiologist such features can have diagnostic significance; see e.g. Bouthan (1968).

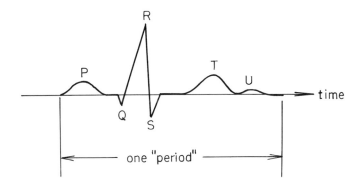

FIGURE 12. PERIOD OF EKG

FIGURE 13. ABNORMAL EKG

The lack of exact periodicity leads one to speculate about a deformation of real physical time into cardiac, subjective time, where the latter is related to the pacemaker activity. A realistic representation of such a deformation will probably be of random nature, and the statistics of the random process will have biological significance.

But other anomalies, such as doubling one of the subgraphs or skipping one, complicates the representation of the process; the subjective time may take a jump, backwards or forwards. It is also necessary to represent events expressed by action on the amplitudes (rather than on time) of the EKG graphs. A segment may be lowered or heightened. This is not the place to deal with this in any detail; let us only emphasize the need for representations not just of typical behavior but also of variations around it.

A class of time recordings that may seem to be completely different from the electro-physiological ones appear in meteorology. If, at a certain location, we record daily averages of temperature or barometric pressure and plot the values over a very long time we get graphs like the one in Figure 14.

Such graphs will of course show seasonal variation if they extend over several years.

28   *Chapter 2: Open Patterns*

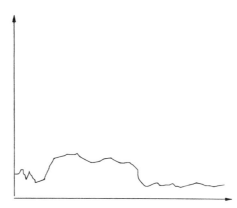

FIGURE 14. REGIME PATTERN

But they can also exhibit a regime-like behavior: over stretches of time, several days, the graphs vary only a little, but then change more drastically. A regime may represent a hot spell, days with prevailing NW winds, and so on.

We cannot avoid at least mentioning secular variation patterns, systematic trends in the development of the climate. This is as controversial as linguistics, with a heated debate that often makes the news. The predictions vary from time to time, sometimes forecasting a downward trend towards an ice age, sometimes an upward trend towards global warming. What they have in common is a promise of dire consequences for mankind delivered by the Cassandras of science.

The behavior of weather regimes is quite different from the approximately periodic EKG patterns. In both cases, however, the underlying physics/biology is understood in considerable detail, but the resulting, overall behavior is still somewhat elusive.

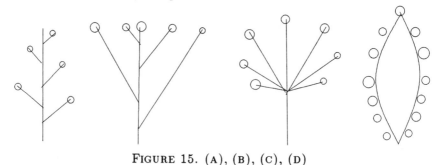

FIGURE 15. (A), (B), (C), (D)

2.7. Plants and trees. We have mentioned graphs of tree type, meaning graphs without any closed cycles. The term "tree" obviously alludes to real, biological trees and the arrangement of branches in them. In Figure 15 we show some different architecture of flowers on a plant. Note how the various parts of the flowers are attached to each other in a systematic way.

Similarly observe the root systems in Figure 16. One can speculate on why these architectures look so different. Perhaps there are different needs for stable rooting in the ground and/or different strategies for extracting nutrition from the soil. Will the attempt to maximize energy input lead to such regular arrangements? A partial, affirmative, answer to this question is given in Grenander (1993), Chapter 6.

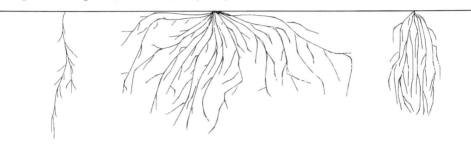

FIGURE 16. (A), (B), (C)

2.8. Tracks. Languages can be thought of as *abstract patterns* since their symbols do not necessarily have any meaning attached to them; recall the discussion of formal languages. Regime patterns describe functions, more generally *contrast patterns* as will be discussed in Part II. In the next chapter we shall consider *set patterns*: objects are sets, and patterns with other types of objects will be encountered later. One important type is the *line pattern*: objects are lines, say, in the plane or in 3-space.

We shall exemplify line patterns by tracks, or trajectories, in the chambers that physicists have been using for a long time to study the behavior, creation, and annihilation of elementary particles (see note 2.4). Such chambers contain a medium that makes the particle tracks visible to some sensor; one observes how particles enter/exit the chamber and other events. A stylized picture of this is given in Figure 17.

Real track pictures, say, from bubble chambers, have a lot of noise in them that will make it harder to analyze them. It will, for example, not always be clear if a track is interrupted by noise or if we are dealing with two tracks.

One could argue against including track patterns together with the other one-dimensional ones in this chapter because they take place in the plane, 2-D, or three-dimensional space, 3-D. This is true but irrelevant. More intrinsic is the fact that a curve, a trajectory, is essentially one-dimensional, although it lives in the plane or space, since it is naturally described by a single real parameter, for example arc length s

$$\begin{cases} x = f(s) \\ y = g(s) \end{cases}$$

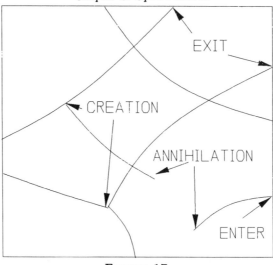

FIGURE 17

After branch points we also need a binary "coordinate" telling us which of the two branches to follow. Therefore these patterns should be considered as one-dimensional irrespective of the fact that they are embedded in a higher dimension. For example, maps can be viewed as one-dimensional and the same applies to road maps, river systems, aircraft trajectories, etc., so long as interest is limited to the curves representing the rivers, roads, and so on. We shall see in Part II that the one-dimensional nature greatly facilitates the analysis and processing of such patterns. This remark is of greater scope than one may think at first. We shall see repeatedly that patterns that seem to have, superficially, some structure are better represented by deeper and more expressive structures. A basic theme in pattern theory is the search for hidden explanations, an activity that could be called *mathematical hermeneutics* (note 2.5).

2.9. Behavior. Behavior patterns seem even less one-dimensional than tracks; they take place in space-time which is three- or four-dimensional. Nevertheless, we shall see that it is sometimes possible to represent them by one-dimensional models.

An instance of this is patterns of domination in animal behavior. Consider a group of individuals from the same species interacting with each other. When herds are formed some of the stronger individuals, perhaps mature stags in a group of deer, tend to dominate the does and weaker stags. Actually this may lead to a hierarchy in the herd, a leader influencing some that in their turn influence others, and so on. On the other hand, two herds may be unrelated to each other in the power structure.

The structure could be represented by diagrams as in Figure 18. There individual 1 dominates 2, 3, and 4, and indirectly 5 and 6, while the leader 7 only dominates 8 and indirectly 9 and 10.

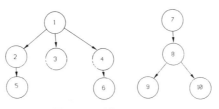

FIGURE 18

Note that the diagram only consists of chains $1 \to 2 \to 5, 1 \to 3, 1 \to 4 \to 6, 7 \to 8 \to 9, 7 \to 8 \to 10$. There are no loops where a chain connects back to an earlier member. Of course this is only a hypothesis; real domination systems could conceivably contain loops. Under the hypothesis, however, the structures are essentially one-dimensional although the diagram displays them in the plane (or plane × time for dynamics).

Another situation occurs in the study of flocking, say of birds or schools of fish. Ornithologists have speculated (note 2.6) over what causes flocking and, more recently over how the behavior should be represented. The question is particularly vexing for species where the flock moves ahead (see Figure 19), but then suddenly and for no apparent reason changes direction. This figure is a computer simulation. Such behavior, that can be quite dramatic, could perhaps be explained in domination terms, postulating the existence of a leader, but this has been argued against by zoologists. What would be reasonable mathematical representations of such behavior patterns?

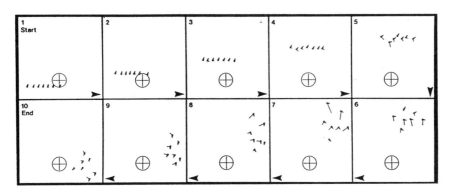

FIGURE 19. SIMULATED BIRD FLOCKING

It is too early to deal with the mathematics of this challenging problem; let us just remark that the circling behavior that is occasionally interrupted by rapid changes of direction is reminiscent of the regime patterns from section 2.6. This suggests that it may be possible to find representations similar to those that we shall study for regime patterns; we shall return to this possibility in Chapter 6.

2.10. Doctrines. Any theory of language, particle tracks, hand forms, and so on, tries to represent knowledge by a formal structure whose constituent parts and rules of combination are chosen as simply as possible without violating the empirical experience that is available. Such representations can be of quantitative form or they may be given in a more qualitative manner.

As an example of the latter let us think of the telling of stories, for example fairy tales. The study of folktales has a long history, but one of the most thought-provoking and original attempts to formalize folktales was presented in 1928 by Vladimir Propp in "Morphology of the Folktale." This work did not really become appreciated in the West until the 1950s. Propp tried to determine the *primitives of folktales*, in his case the stories in Afanasev's "Russian Folktales." Here are a few of them:

$$\alpha = \text{initial situation}$$
$$\beta_1 = \text{departure of elders}$$
$$\beta_2 = \text{death of parents}$$
$$\ldots$$
$$A^1 = \text{kidnapping of a person}$$
$$A^2 = \text{seizure of a magical agent}$$
$$\ldots$$
$$B^1 = \text{call for help}$$
$$D^1 = \text{test of hero}$$
$$E^1 = \text{sustained ordeal}$$
$$\ldots$$

all together a couple of hundred basic events.

He then *combines* them into strings, for example as for the following story:

A tsar, three daughters (α). The daughters go walking (β^3), overstay in the garden (δ^1). A dragon kidnaps them (A^1). A call for aid (B^2). Quest for three heroes ($C \uparrow$). Three battles with the dragon ($H^1 - I^2$), rescue of the maidens (K^4). Return ($\downarrow$), reward (w^c). The formal description could then be the representation

$$\alpha \ \beta^3 \ \delta^1 \ A^1 \ B^2 \ C \uparrow \ H^1 - I^2 \ K^4 \ \downarrow \ w^c$$

Arbitrary strings over the alphabet $\{\alpha, \beta_1, \beta_2, \ldots\}$ may not make much sense; *the elements must follow each other in some reasonable order*. Propp handles this issue by diagrams that describe how substories follow each other. For example, action may be interrupted by an episode, and after the episode is completed the main story continues:

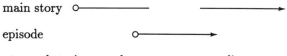

Or, two substories may have a common ending:

story I

story II

The ideas that were introduced in Propp's seminal work are of general scope. Other doctrines that attempt to formalize a given ensemble of actions and behavior may be analyzed in the same spirit, breaking down each observed event into fundamental subevent. Such a *combinatory approach* could be attempted for various disciplines. For example, statistical doctrines of today (or one of them) could conceivably be expressed in terms of a set of actions/principles, a set that would perhaps not be very large. A reader with a speculative mind can no doubt come up with other potential applications of Propp's ideas.

2.11. Open patterns. By now it may have become clear to the reader why we have grouped all the patterns of this chapter together. Their structure shares a one-dimensional flavor, not in the sense of dimension in geometry, but in terms of the dependencies, the logical couplings in their structure. They are *linear* arrangements, one part following another. There are no closed loops in their topology that will later on be described via graphs: they are *open*, meaning the absence of cycles.

Chapter 3

Closed Patterns

In contrast to the open patterns in our catalog the closed ones can possess intricately woven systems of dependencies. This induces a topology, an *information architecture*, that is often hidden from the observer. The pattern analyst faces a serious challenge in finding meaningful representations for closed patterns since their surface appearance may not give a clear indication of what deep (regular) structure supports them. Their mathematical treatment (Chapter 6) will require more thought.

3.1. More ornaments. It is best, however, to start with some obvious cases extending section 2.1 into two dimensions. The ornament in Figure 1 is straightforward and its structure needs no explanation.

FIGURE 1. BYZANTINE ORNAMENT

On the other hand the pattern in Figure 2 has a more intricate structure where the regularities are not so easy to describe in words but still clearly visible. These illustrations are from Owen's "A Grammar of Ornaments" referred to earlier. The "grammatical" analyses that accompany the ornaments in his work are grammatical in a different sense

Chapter 3: Closed Patterns 35

FIGURE 2. A MORE COMPLICATED ORNAMENT

from that used in section 2.1, but some of Owen's ideas will sound familar to a reader knowledgeable in pattern theory.

But now some patterns with a less obvious construction. First look at Figure 3.

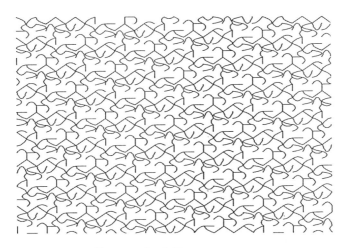

FIGURE 3. 2-D ORNAMENT

It is not easy to see how it has been formed, what are the elements that generate it, and how have they been combined. But in Figure 4 we show a detail of a similar ornament blown up in order to make it easier to understand the structure. Some curves seem to appear repeatedly, and closer inspection of a sample of such pictures will show that these elements are the ones in Figure 5 together with their rotations by 90°, 180°, 270°.

We can now code Figure 3 into the alphabet 1,2,3,4,5,6,7 and get an array of numbers. The entries in this array appear to be, and are, purely random.

FIGURE 4. DETAIL OF ORNAMENT

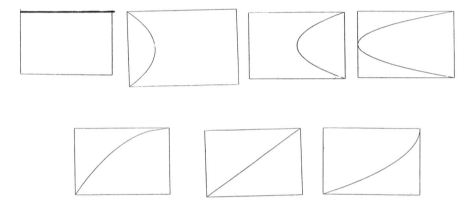

FIGURE 5. ELEMENTS OF ORNAMENT

After this we are ready for some more structured patterns over the same alphabet. Look at Figure 6, which looks more regular with clustering occurring in four places. Obviously the picture elements influence each other, but how?

To understand such patterns we need some coordinate system to determine what element should appear where in the image. "Coordinate system" must be taken in a wide sense – the "coordinate axes" need not be at right angles to each other, and the coordinate pair (x, y) may refer to more than one point in the image. But is this really necessary? To determine what element should be at location (x, y) is it not enough to know the elements in the *environment* of (x, y), that is the ones at the location $(x-1, y), (x, y+1), (x+1, y), (x, y-1)$? Figure 6 is based on such a rule, but it is difficult to see this without a detailed analysis of the coded picture.

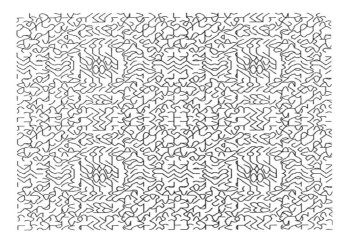

FIGURE 6. ANOTHER 2-D ORNAMENT

In section 2.1 we discussed (in one dimension) such purely local constraints, where no absolute coordinate system was used, only information about *relative positions*, and this immediately extends to two dimensions. We shall return to this question many times and make it more precise in Part II.

Such patterns are more rigid; they can have little or no variability. A different behavior is shown in Figure 7, which has some randomness in it. Note, however, that it is far from being chaotic. A's tend to be surrounded by A's, B's by B's. Here we have an instance of something that often occurs in real life patterns: *controlled randomness*. But how does one represent and analyze such patterns? This difficult question will be studied in Part II.

The strategy is now becoming clearer: *first determine what are the elements, the building blocks, that make up the images, then find how the elements are coupled together.* To do this we need not only intuition and subject matter knowledge, but also analytical and computational tools such as the sophisticated techniques to be introduced in Parts II and III.

In simple situations it may of course be possible to just guess the elements. Look at Plate 3, where the images have been generated by fairly simple rules but the result is hidden by some limited randomness. This looks like a jigsaw puzzle and that is just what it is. Pieces are joined together, the pieces are quadrilaterals with curved and jagged boundaries, and two pieces are joined only if the boundaries fit. This is more than a puzzle, it resembles *conformation* (note 3.1) of molecules in chemistry. The rules say that in some sense the primitives, here meaning either quadrilaterals or molecules, must conform in some yet to be made precise way.

$$ABBBBBAABABBBBAB$$
$$AAABBABBABBBBBAB$$
$$AAAAABBBABBBAAAA$$
$$AAAABBAAAABABBAB$$
$$BABABBBAABBABBBA$$
$$AAAABAABBBAABABB$$
$$BBBBABAABAAAABA$$
$$BAABBABABABBBABA$$
$$ABBABGAAAHABBBAB$$
$$AABABBEBFBBBAABB$$
$$ABABBAADAAABBAAB$$
$$AABABAEBFABAAABA$$
$$BBAAAGBABHAAAAAA$$
$$ABBBABABBAAAAAAB$$
$$AAABAAABBABBBAAA$$
$$BBABBABBABABABBB$$

FIGURE 7. CONTROLLED RANDOMNESS

Now look at Plate 4, which shows pictures of jigsaw puzzles "grown" from generators, starting with one generator (curved quadrilateral) selected at random from the given set. The algorithm then looks (at random) for some new generator that fits along one of the four boundary arcs, and so on until no more fit is possible. It is similar to Plate 3 but the puzzle becomes saturated before the whole square is filled with generators. This is because the set of available generators is larger; some do not fit each other at all.

3.2. Weaving. Weaving patterns can be produced by many sorts of looms, of which the following is a typical example. This loom consists of five parts:

1. The warp beam around which the *warp* yarns are wound.

2. The *harnesses* with heddles through which the warp yarn has to pass. Each harness frame can be raised or lowered to let the *filling* yarn pass over or under a group of warp yarns.

Chapter 3: Closed Patterns 39

3. The shuttle containing the filling yarn. The shuttle passes from one side of the loom to another. The filling is also call the *weft*.

4. The reed looking like a big comb. All the warp yarns pass through it, and its function is to push the filling yarn against the already formed fabric.

5. The cloth beam on which the new cloth is taken up.

Let us denote the number of harnesses by r, and the number of filler yarns, perhaps of different color, by s. The weaving patterns are practically (but not exactly) two-dimensional. We will show the simpler ones in black and white but they can be much more colorful in reality; some color weaves will be displayed later.

Let us look at some of the simpler weaving patterns. With two harnesses only, $r = 2$, and one filler, $s = 1$, one can produce the *simple weave pattern* in Figure 8 periodically. The warp yarns pass through harness 1 and 2 in an alternating manner. Harness $h_1 = \{1, 3, 5, \dots\}$, $h_2 = \{2, 4, 6, \dots\}$.

FIGURE 8. SIMPLE WEAVE

A simple variation of this is the *basket weave* of which an example is shown in Figure 9. The filler consists of two yarns, a double pick, and the plane pattern has a characteristic checkerboard appearance. Here $h_1 = \{1, 2, 5, 6, \dots\}$ and $h_2 = \{3, 4, 7, 8, \dots\}$ and the filler is a double yarn.

FIGURE 9. BASKET WEAVE

The *twill weave* that needs at least three harnesses (shafts) is illustrated by Figure 10. It is called a three-shaft twill and the produced plane pattern has the diagonals typical for

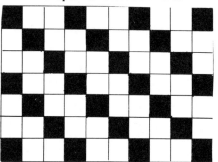

FIGURE 10. TWILL WEAVE

twills. More complicated shapes can be produced by varying the diagonal through changes in the warp threading. Here $h_1 = \{1, 4, 7, \ldots\}$, $h_2 = \{2, 5, 8, \ldots\}$ and $h_3 = \{3, 6, 9, \ldots\}$.

One such variation is shown in Figure 11. It employs four harnesses in which the threading order is given by the difference scheme $\ldots 5, 3, 5, 3, 5, 3, \ldots$, so that starting with warp yarn no. 1 we get $1, 6, 9, 14, 17 \ldots$. This gives us the first harness, h_1, and so on for the others. The space-time configuration has a cycle as shown in the figure with the harnesses changed by adding 1 to its number modulo 4. These motions produce the *herringbone patterns* in the figure.

FIGURE 11. HERRINGBONE WEAVE

For *satin weaves*, finally, one needs at least five harnesses; a smaller number would only give twill patterns. An example of a five harness satin is shown in Figure 12 where the warp threading is given by the difference scheme $\ldots 5, 5, 5, \ldots$, so that starting with warp yarn no. 1 we get the first harness $1, 6, 11, 16, \ldots$ and so on. The time pattern for successive harnesses is given by adding 3 modulo 5 to their number. The result will depend upon whether the surface is dominated by the warp – warp satin – or by the weft – weft satin.

The plane pattern produced is a weft satin. Note the long floats typical of satins which tend to reflect light and give the surface its luster.

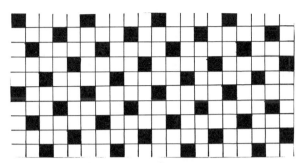

FIGURE 12. SATIN WEAVE

How can we represent these and similar patterns formally? One way is as follows. Note that all of the five figures are periodic. Take for example the twill in Figure 10 and consider the rectangular piece of it that consists of the first three rows and three first columns. This piece repeats itself horizontally and vertically. To get a local description let us introduce a coordinate system (x, y) as in Figure 13 so that both coordinates repeat the cycle $(0, 1, 2)$. A square with coordinates (x, y) is colored black if $x = y$ and white otherwise.

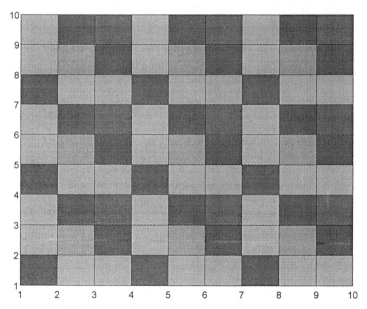

FIGURE 13. LOCAL COORDINATE SYSTEM

Let us make a seemingly trivial observation. Consider a square with vertical coordinate x. Its neighbor in the downward direction will then have the vertical coordinate $x + 1$ if $x = 0$ or 1, and $x = 0$ if $x = 2$. Formally, use the ternary number system over the numbers $\{0, 1, 2\}$ and with the addition table

+	0	1	2
0	0	1	2
1	1	2	0
2	2	0	1

In other words, consider all integers reduced modulo 3 by division, keeping the remainder only.

With this set up we have a *local coordinate system* in the sense that we can compute the coordinates (relative!) of a square if we know one of its neighbors, for example downwards we get $x + 1$ from a square with vertical coordinate x.

Also, if a square has the local coordinates (x, y) we can compute its color in B/W: black if and only if $x = y$. On the other hand the knowledge that a square is, say, colored black does not determine x or y. Also, the knowledge of (x, y) certainly does not determine the absolute position of the square.

This representation is adequate for periodic weaves (others will be introduced in section 8.7) but can also be applied to other patterns that superficially seem to have nothing in common with weaving. We are thinking especially about *crystals*.

In Figure 14 we show a crystal displaying the sodium and chlorine atoms as they are combined together in a regular arrangement.

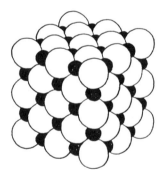

FIGURE 14. SODIUM CHLORIDE CRYSTAL

Real crystals are three-dimensional but for didactic reasons let us look at them in the plane only. In Figure 15a is shown a regular arrangement of three types of atoms denoted

·, ×, and o. The plane is divided into congruent parallelograms, fundamental cells, one of which is isolated in 15b. In 15a there are also shown two (oblique) coordinate axes x and y. Any point in the plane can be written as a linear combination $x\xi + y\eta$ of the two basis vectors ξ and η in Figure 15b. This is just as for an ordinary (rectangular) coordinate system in analytic geometry.

The ×-atoms are located at the points $n\xi + m\eta$, where n and m are arbitrary integers. Then we can write for the location of an arbitrary o-atom $(n+\alpha)\xi + (m+\beta)\eta$ where α and β are fixed real numbers between 0 and 1. Similarly the ··-atoms are at $(n+\gamma)\xi + (m+\delta)\eta$; $0 < \gamma < 1$, $0 < \delta < 1$.

But the $\alpha, \beta, \gamma,$ and δ system is a direct extension of the local coordinates used for periodic weaves. Note that if we translate the whole plane a whole number n of ξ-vectors in the x-direction and a whole number m of η-vectors in the y-direction the (infinite) crystal looks the same. We say that the crystal is *invariant* with respect to these movements of the plane. Similarly, a crystal may be invariant with respect to some rotations around a point. In general one can characterize a crystal partially by specifying the *group of movements* that leave it invariant. This is the basis for classification in crystallography. The reader will find an elegant account of this and related topics in Weyl (1952).

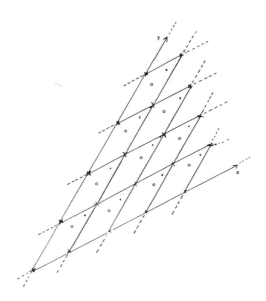

FIGURE 15A. CRYSTAL LATTICE

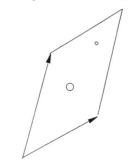

FIGURE 15B. FUNDAMENTAL CELL

3.3. Textures. We now turn to patterns less regular than the periodic ones and begin with the one in Figure 16. It is almost immediately seen that it consists of ellipses with varying location, size and orientation, some very small. There seems to be no obvious regularity in the way these parameters vary, and this is indeed correct; the picture was synthesized with random locations, sizes, and orientations.

FIGURE 16. ELLIPSE PATTERN

In Figure 17 it is also easy to see that the generators are the symbols ∗ and o, but now the elements are not placed at random. Instead, the ∗-symbols are clustering around the o's so that the elements interact with each other; they are not purely random as was the case in Figure 16.

The interactions are not as systematic as in the patterns in section 3.2, however, they are partially random. Models of this type, where the centers are randomly placed, and then secondary elements attracted to the centers, have been used in geography to represent economic patterns. The centers then mean some resource and they are surrounded

Chapter 3: Closed Patterns 45

FIGURE 17. CLUSTERING PATTERN

by elements representing some economic activity. An example of clustering of inhabited settlements is given in Figure 18. Another example is from astronomy: clustering of galaxies. Many other applications of clustering patterns are in use (our galaxy, the Milky Way, belongs to a cluster of 15).

FIGURE 18. CLUSTERING OF SETTLEMENTS

One meets patterns with similar *logical architecture* but with different elements in the study of surface appearance of various materials. So, for example, do we get pictures of cloth in Figure 19 (note 3.2), showing the systematic, slightly perturbed, arrangements of fibers along two perpendicular directions. The fur texture in Figure 20 shows fibers in continuously changing directions, while the wood grain in Figure 21 has a family of fibers from annual rings of the tree.

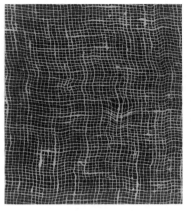  

FIGURE 19. CLOTH WITH FIBERS   FIGURE 20. FUR   FIGURE 21. WOOD GRAIN

These patterns also have invariances, not strictly deterministic ones as for crystals, but in some average (statistical) sense. In Figures 16 and 17, for example, if we translate or rotate the images they will still possess the same average couplings. In Figure 19, on the other hand, the same can be said for translations but not for rotations; the direction of the fibers distinguishes one direction from the others.

In Figure 22 we show an image of elements that are straight line segments and whose location and orientation have statistical invariance with respect to rigid motions consisting of translations and rotations.

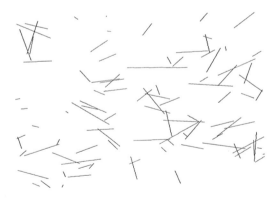

FIGURE 22. ISOTROPIC LINE PATTERN

In Figure 23, on the other hand, we do not have statistical invariance with respect to rotations since there is a preferred orientation in the vertical direction.

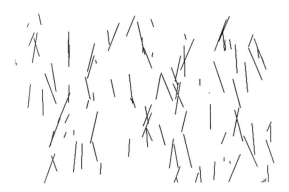

FIGURE 23. ORIENTED LINE PATTERNS

When the group of translation movements leaves the statistics unchanged we could say, with some exaggeration, that nothing really happens in the picture, it is essentially the same everywhere (what should that really mean?). Such patterns we shall call *textures*, but this definition is not universally accepted.

When we say that the image is statistically the same as a translated version of it we should, strictly speaking, deal with infinite picture in the whole plane. Otherwise the statement is hard to interpret, but practically we will be satisfied if the picture is so large that the coupling between distant elements is weak. Also we would like to have access to an *image ensemble*: a (large) collection of pictures that are in some sense related. If we do, then it is possible to speak of statistical properties with more confidence. This is too vague and we shall tighten up the definition and discussion of textures in Chapter 6.

Texture patterns may not be very exciting but are important in that for a given picture some of its subpictures present different textures to the observer. An example in Figure 24 consists of an aerial photograph taken in the North Atlantic of the sea. One can distinguish between areas of ice and open water. The picture is acquired by synthetic aperture radar (note 3.3).

The whole image has been broken up into segments of different textures. An important practical problem in image analysis is to automate texture segmentation. To achieve this we need some way of representing texture patterns mathematically and we shall return to this task in Part II. Let us remark that textures will later on be used as building blocks for representing more challenging image ensembles.

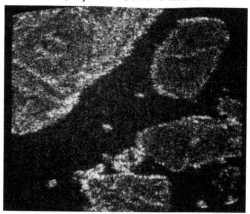

FIGURE 24. HETEROGENEOUS TEXTURE PATTERN

3.4. Shapes. Instead of the homogeneous, "eventless" textures we shall now examine pictures where something really happens, some areas stand out against a background. To be concrete, let us consider an ensemble of pictures where each picture represents a set and all these sets have some regularity and resemble each other. For example, in the stomach shapes in Figure 25 each picture consists of a boundary, a closed curve that contains a set, its interior. This picture is from Anson (1963), an anatomical atlas that is of particular interest for us here. Not only does the atlas present normal and abnormal specimens – many anatomical treatises do that – but it also attaches numerical values, frequencies, to the variations. Thus the user of the atlas is given some idea of how likely or unlikely are the different variations of an anatomical specimen. The stomachs in Figure 25 are to be understood as two-dimensional projections of the three-dimensional stomach shapes. *Set patterns* like these can be described simply by a contour; no information is given about the inside or outside but we shall return to this topic in the next section.

Look at the left upper stomach. Its contour has a long, convex arc, the *curvatura major*, at the bottom and right side, and a shorter, concave arc, *the curvatura minor*, at the top and left. It also has two openings. The other stomachs also have these features to a varying degree and it is clear that they belong together although it is not always clear where, for example, the *curvatura major* is. They are deformed versions of the first one.

Suppose that we take a picture of the first stomach with an idealized digital camera of resolution $H \times V$, meaning that for each integer x between 0 and $H-1$ and each integer y between 0 and $V-1$ we get a value denoted $I(x,y)$. Say that $I(x,y) = 1$ inside the stomach and $I(x,y) = 0$ outside where x,y determines a little picture element, a pixel, whose left lower corner has coordinates x and y. If we interpret the value 1 as black and 0 as white we will get a figure like the one in Figure 26. It is very crude since the resolution is only 11×13 and since the image values are only 2, 0 and 1. Real digital cameras have resolutions of the order 512×512 or more, with at least 256 gray levels, or perhaps color values.

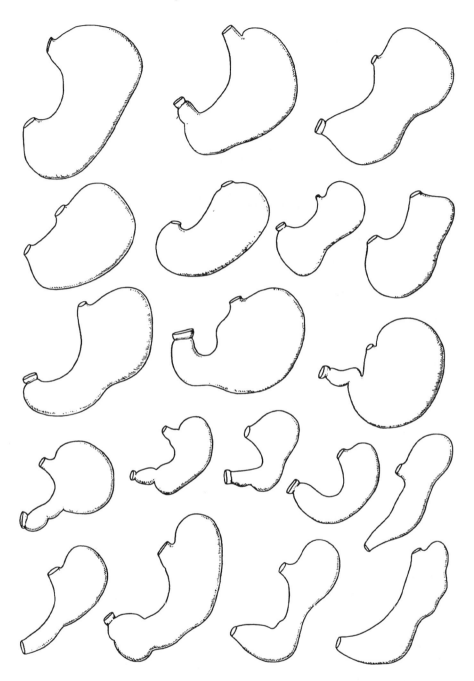

FIGURE 25. STOMACH: VARIATIONS IN FORM

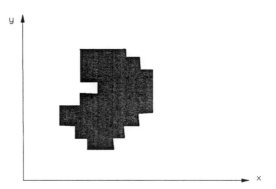

FIGURE 26. DIGITAL PICTURE

To be able to compare the different pictures in an ensemble numerically with each other it is useful to standardize them. This can mean many things:

a) Translate the set in the picture to standard position, for example by moving it so that its center of gravity coincides with the midpoint of the picture;

b) Rotate the picture, for example so that its main direction (perhaps using the major axis of the moments of inertia, note 3.4) points upwards;

c) Scale the picture, for example so that the area of the set becomes half of the whole picture.

For textures such standardization is not very useful since we have no "handles" in the picture to know where things are, but shape patterns have those handles, as will be seen.

In Figure 27 we show set patterns formed by the boundaries of leaves from trees: (a) an oak leaf, (b) a maple leaf, and (c) a sassafras leaf.

These pictures have been produced as follows. Leaves are collected, dried and pressed flat, then placed on top of a light table or a small glass plate. Light comes up through the plate. A digital camera is placed over the plate in such a way that its height is easy to move and focus the camera. Using some imaging software (in the case of Figure 27 it was a package called IMAGEPRO, a registered trademark of AT&T), the digital picture is captured and stored on the hard disk of a PC. The resolution was 512 × 480. Some standardization was

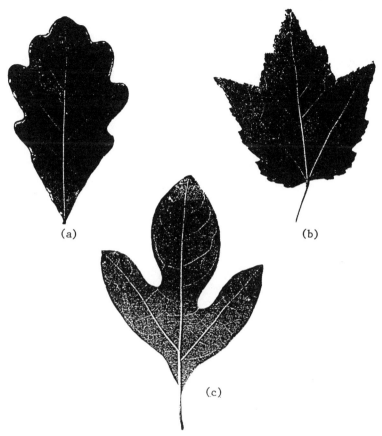

FIGURE 27. LEAF SHAPES

achieved by putting the point on the leaf boundary where the stem enters the leaf at a given point on the light table and with the main axis of the leaf pointing along the vertical axis of the light table (note 3.5).

Leaf shape is highly variable, even within the same species. In Figure 28 we show four sassafras leaves, each with the three characteristic lobes but with many differences in the detailed behavior of the boundary. Also topology – number and arrangement of lobes – can vary more than shown in the picture.

The boundaries form closed curves so that it is natural to represent them (approximately) as polygons made up of *generators* consisting of line segments joined together continuously as in Figure 29. Each element has two neighbors that we could call "left" and "right." Note, however, that in contrast to what was the case in Chapter 2, the elements are bound together in a closed loop – which is why we classify them as *closed patterns*.

FIGURE 28. SASSAFRAS LEAVES

Another family of biological pictures that will be analyzed in Chapter 6 consists of human hands. They have been imaged as the leaf pictures except that the optical conditions were made so poor (intentionally) that the boundaries are not always clearly visible. They are from adult males who were asked to put their right hand pressed down against the light table and with the left side of the wrist at a certain mark. No other instructions were given; for example, the fingers could be held together or spread out. Figure 30(a) shows one such picture and Figure 30(b) shows the contour from a hand captured with good light conditions.

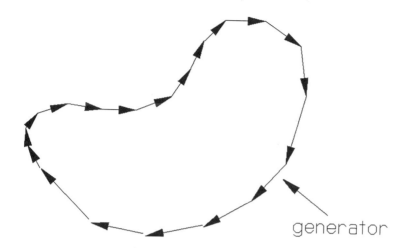

FIGURE 29. DISCRETIZED BOUNDARY

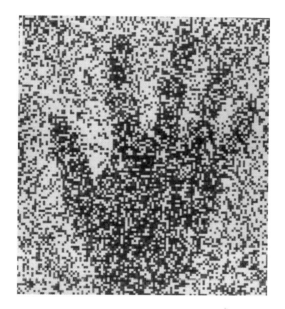

FIGURE 30(A). NOISY PICTURE OF A HAND

Hand pictures are quite characteristic: five fingers (for normal hands) with a shorter

54   *Chapter 3: Closed Patterns*

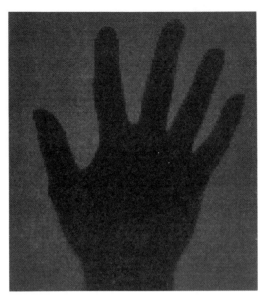

FIGURE 30(B). CLEAR PICTURE OF A HAND

thumb, the length of the other fingers longer, longer, shorter, shorter. The shape of the individual fingers is also fairly characteristic, tapering towards the top except for a widening at the first joint. At the same time they vary a good deal. See Figure 31, where we have shown four hands. The size differs a good deal from one individual to another, for example the width of the fingers and the form of the top, more or less pointed. The ratio width/length of the hand also shows considerable variation. Even for the same individual, pictures taken at different times can differ substantially from each other; the angles between fingers has a range between $0°$ and $45°$ or so.

How could one represent hand shapes formally? Given one or several pictures from the same individual, one could measure the area of the whole hand, of the palm area, the length of the thumb, and other features of this kind. A moment's reflection shows, however, that they are not well defined. For example, how would we define precisely the length of the index finger. One way is to introduce *landmarks* (note 3.6) as the dots in Figure 31. For the moment let us avoid the question of how to define and choose the landmarks, and only emphasize that this whole discussion is too vague and needs to be tightened up.

The ubiquitous opposition "typical shape – variability" has been a major theme among researchers dealing with biological shapes and has generated a massive literature. One celebrated work that should be mentioned before all others is d'Arcy Thompson's *On Growth and Forms* (see References). This monumental work is filled with both beautiful illustrations, most of them drawings, and thought-provoking remarks. This is true especially for the last chapter, entitled "On the Theory of Transformations, or the Comparison of Related Forms," where Thompson suggests that some of the shape changes during evolution can be described by transforming curvilinear coordinate system. The simplest way to explain this is to refer the reader to the drawings in Figure 32, which show how points in (a) with coordinates from the set $\{0, 1, 2, 3, 4, 5\}$ and $\{0, a, b, c, d\}$ correspond to "homologous" points in the two lower pictures. It goes without saying that just as this device can describe interspecies variation (between different species) one could also use it for intraspecies variation (within a given species). It is less clear, however, how this view of shape can be made mathematically precise with numerical characterization of the amount of variability; this topic will be treated in Part II.

Biological patterns will be our main interest here, but one could elucidate the structure of man-made patterns from the same perspective. The difference will be quantitative rather than qualitative. Indeed, think of the shape of automobiles. How much variability can we expect for a given automobile? Well, first there is location, two coordinates, then rotation, one more parameter. If it is a four-door model we need one angle per door, similarly two angles for the lids to the trunk and the hood. That means nine degrees of freedom and there may be others.

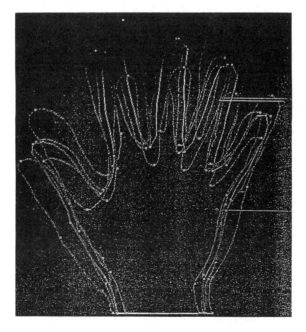

FIGURE 31. HAND WITH LANDMARKS

This assumes that the picture was captured by an ordinary visible light camera. If instead we use an infrared (IR) camera we will need many more parameters to describe the temperature signature. For example, if the engine has been running it will show up as a hot spot in the picture. Leaving aside what to do with IR pictures it is clear that a particular automobile is fairly well described by a small number of parameters. This amount of variability is small compared to the numbers needed for most biological patterns.

This is for intra-pattern variation. If we allow automobiles of several makes, models, colors, and so on, we have inter-pattern variation in addition. Nevertheless, the total number of parameters will remain moderate (but large).

The same is true for many manufactured patterns; they tend to exhibit less variability than the natural ones, probably due to the technological constraints of the manufacturing process. When this is true they should be easier to understand and represent than patterns in nature.

3.5. Inside structure. The hand pictures in the last section consisted just of boundaries, and said nothing about skin texture, color, etc. If we replace the visible light camera as a sensor by an X-ray camera much more internal information is obtained: the bones that make up the hand, cartilage that binds them together, the more transparent soft tissue. We get pictures like the one in Figure 33.

Instead of the simple pattern representations in Figure 29, last section, by just a contour, a simple curve, we need a more detailed one. While the earlier ones could be said to be one-dimensional representations (in terms of the boundaries) of two-dimensional images we now need truly two-dimensional ones. How can this be done in a way that catches some of the underlying biological mechanism?

It is instructive to look at several pictures like Figure 33. One observes immediately that for normal hands two pictures can be made to correspond pointwise to each other. In other words if $z_1 = (x_1, y_1)$ is a point in the first picture we can find a point $z_2 = (x_2, y_2)$ in the second one in such a way that the *correspondence* $z_1 \longleftrightarrow z_2$ associates *homologous points* to each other. If z_1 is some landmark in picture no. 1, then z_2 is the (same) landmark for picture no. 2. We apologize to the reader for being vague here, that will be remedied in Chapter 6.

Chapter 3: Closed Patterns

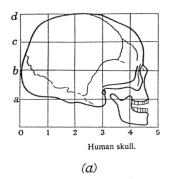

Human skull.

(a)

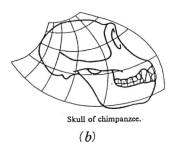

Skull of chimpanzee.

(b)

Skull of baboon.

(c)

FIGURE 32. TRANSFORMED IMAGES

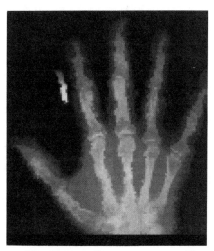

FIGURE 33. CLEAR PICTURE FROM X-RAYS

A mapping $h : z_1 \longleftrightarrow z_2$ such that to each z_1 corresponds one and only one z_2, and vice versa and such that z_2 is a continuous function of z_1, and vice versa, is called a *homeomorphism*. With this terminology we could express the above by saying that we represent an image ensemble of X-ray pictures of normal hands as a homeomorphic map hI_{temp} of a *template picture* I_{temp}.

But have we gained anything by such a representation? To answer this question let us go back to the contour-based representation of set patterns in section 3.4. Say that that resolution is $H = V = 512$ and that we use $n = 128$ line segments to represent the set. The whole picture has $512 \times 512 \cong 1/4$ million pixels and since we deal only with B/W pictures one bit per pixel is enough when we store the picture on the hard disk. Since 1 byte = 8 bits we need approximately 30,000 bytes for the picture. For the boundary we use n x–coordinates and n y–coordinates. If we store each coordinate as integers we can use 4 bytes per coordinate, altogether $4 \times 2 \times 128 = 1024$ bytes. The representation therefore achieves the information compression $30,000/1024 \cong 30$, a considerable gain in the storage requirement.

For the X-ray picture we need 4 bytes per pixel, which is about 1 million bytes if we store the intensities as integers (1/4 of that gives acceptable approximation if we use only 256 gray levels). This is stored once and for all only for the template. How much storage do we need for the mapping h? Strictly speaking the mapping function h ought to be given in floating point arithmetic, say with 8 bytes for value, so that we would need $8 \times 1/4$ million = 2 million bytes. If we use only integer arithmetic, we get 1 million bytes. Whatever we choose we see that, disappointingly, no information compression is obtained. Remember that a picture in the ensemble is determined by h; the information in the template is stored once and for all.

But this is misleading. Indeed, for any practical purpose we do not need all the details in the mapping function h. As we shall see in Chapter 6 we will get excellent results with only a fraction of the information in the mapping function h. Exactly how this should be done will be left till later; suffice it to say that around 100 floating point numbers, 800 bytes is sometimes enough. That leads to information compression of about 1000 for h (but not for the template).

Two qualifications are advisable here. First, the main role of pattern representations in the present context is not to achieve information compression. If that were the primary objective, much higher compression rates could be achieved. Instead, the representations are intended to serve as tools for pattern understanding/processing, as we shall explain later on. We gain knowledge. Second, let us not forget all the information that has to be stored for the template. Although this is static storage, done once and for all, it takes up space.

The mathematical modeling in mechanics and physics has usually been achieved by including a few constants, parameters, in the models: mass of the electron, speed of light, the gravitational constant, and so on. Biological patterns require massive numbers of parameters, not to characterize a particular image but for a whole image ensemble. We mentioned something like 1 megabyte for the template. Such models could be called *megabyte models*.

One could speculate that during evolution so many chance factors have influenced the formation of a species that it would be implausible that the resulting shape, say of an internal organ, ultimately expressed by a DNA sequence, could be described by a mathematical expression with a small number of parameters. Whatever one may think of such speculations, it appears that megabyte models will be forced upon us when we attempt to represent increasingly complex biological/medical patterns, say, in 3-D and with many sensor modalities, and if so, it will force us to develop mathematics adequate for creating and analyzing such models as well as software implementing the mathematics.

3.6. Connections. We are now beginning to see an emerging theme in pattern representations. When we form patterns from elements to depend upon each other, they are coupled together, perhaps by deterministic relations, perhaps by random arrangements, by a logical architecture of *graph type* (note 3.7).

The way these couplings extend over the whole pattern can vary a good deal. In Figure 34 the simplest arrangements are in (a) and (b); in (a) we show a linear type of connection, in (b) a tree type. Such arrangements have no loops – they describe open patterns. In (c) and (d) loops are allowed (closed patterns): in (c) we have a cyclic arrangement and in (d) a square lattice design (undirected segments). These diagrams show the *information*

60   *Chapter 3: Closed Patterns*

architecture of the patterns: how one element directly influences some other elements. An element can therefore be said to have some others as *neighbors*. Only these are directly influenced by the first one, the others may also be influenced, but only indirectly, via some other intervening elements.

We shall now discuss some such architectures, but first we must consider how an observer acquires the information. The first possibility one might think of is the way humans and most animals visualize the world around them. Light in the range of visible light is reflected from, or sometimes emitted by, the environment so that variable color and intensity is registered from different directions. In addition, depth information is gained from the binocular vision achieved by two eyes cooperating and by movements of the head.

How visible light is reflected from opaque objects or transmitted through partially transparent ones is a complicated business. In theory it should be easy to understand, using long established principles of optics, but it is really very difficult. Reflection depends upon, for example, the nature of the surface against which it is reflected. It could be specular reflection, as against an ideal mirror, or Lambertian reflection, as against a rough surface, or it could be something intermediate. Shadows and obscuration complicate the problem even further.

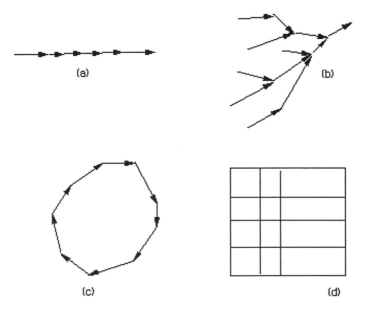

FIGURE 34. CONNECTOR GRAPHS

Whatever the case may be, once the light reaches the retina the physiology of the immediate processing, cones and rods and so on, is understood in great detail; a large body of knowledge has accumulated during this century using many disciplines, in particular electrophysiology. Processing during the later stages, higher level vision, is much less understood, in particular the global, overall processing that enable us to recognize objects and in general to understand the picture (see note 3.9).

But this is only one type of vision. The anatomist or histologist who wishes to examine the fine structure of some tissue has long been slicing the object by a microtome, perhaps staining the slice, and viewing the preparation in a visible light microscope. This technique, which goes back more than a hundred years, suffers from the disadvantage that the organism is dead when viewed. A recent modification, optical sectioning, avoids the physical slicing and allows the organism to be alive so that one can see, for example, motion. A radiologist can illuminate the scene, not by visible light, but by X-rays or some other radiation. Whatever the particular technology, it is clear that we lose information when we get 2-D representation of 3-D scenes. A fundamental problem is therefore how, and to what extent, we can recover 3-D knowledge from one or several 2-D picture. A whole scientific discipline, *stereology* (note 3.10), has grown up centered around this question. The stereologist would try to extract 3-D information about the spatial structure that results in such a slice.

In everyday life we are familiar with many artifacts that simulate human vision. An example is the ordinary camera, in black/white or color, another is the movie camera and camcorder. With electronic technology we get digital cameras where the output is a matrix of intensities for some wavelengths of the light.

Computer vision is the technology that tries to automate the vision process by technological artifacts. It is tempting to build devices that simulate the human vision process to the extent that it is known. After all, human vision is extremely powerful, fast and highly precise. But nothing prevents us from basing the construction on other principles. One such device is the *laser radar* which, at least in principle, collects range data: in each direction it measures the distance to the nearest object. That means that the output is purely geometric, the surfaces that bound the objects in the scene are represented in spatial polar coordinates; distance as a function of the two space angles up/down, left/right. The actual functioning of the laser radar is much more complicated than this idealized description, and the accuracy is not overwhelming but that will probably change for the better soon.

A final example of how 3-D scenes are viewed by devices that give 2-D output is in the medical imaging technology known as *tomography*. Tomography comes in many versions depending upon what type of radiation is used, but the basic idea is as follows (note 3.8). Radiation of some sort is captured by a recording device (camera) with an array of sensors (see Figure 35). The array can be rotated by an incremental angle a full turn. The resulting image will have little obvious relation to the object being scanned (a head in the figure); instead it has a typically wavy form, the sinogram in Figure 36, which has to be analyzed in order to give useful biomedical knowledge. Many tomographic algorithms have been invented for this purpose. What is observed can often be expressed as a line integral, the line intersecting the object.

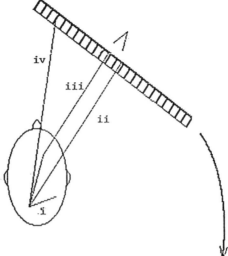

FIGURE 35. TOMOGRAPHY SETUP

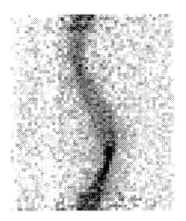

FIGURE 36. SINOGRAM

After this technological aside let us return to what is our main concern here – what sort of 3-D patterns do we expect to encounter and how will their elements interact with each other? The stomach shapes of section 3.4 were 2-D projections, but real stomachs live in 3-D. We get a clue from the regime patterns in Chapter 2 that were represented as combinations of simple curve elements.

For the surface of a stomach shape we could try to represent it as a combination of simple area elements, say flat pieces. In other words, the surface would be approximated by a polyhedron, as is shown in Figure 37.

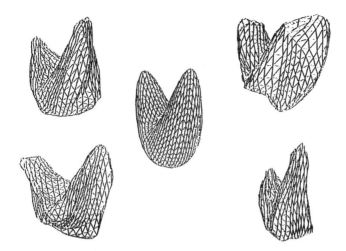

FIGURE 37. WIRE FRAME STOMACHS

But which polyhedron should we choose? This is not obvious, not even for the simplest surface, the sphere. When we describe a curve by a curve I by a linear spline I_s as in Figure 38, it seems reasonable to choose the dividing points on the x-axis, the so-called knots, evenly distributed, so let us do that on the sphere too.

Evenly spaced points on the sphere – this statement is a bit vague. On a circle it is clear, we can rotate the circle by an angle so that the set of knots moves to coincide with the set itself. On the sphere we can do something similar, for example select the points that correspond to the inscribed tetrahedron. But that construction gives only four points. The inscribed cube gives eight points and so on, as in Figure 39 and summarized in Table 1.

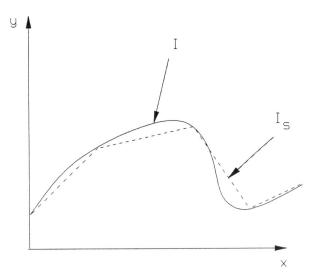

FIGURE 38. SPLINE APPROXIMATION

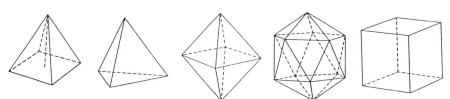

FIGURE 39. PLATONIC SOLIDS

Table 1			
	no. of vertices	no. of segments	no. of flats
tetrahedron	4	6	4
cube	8	12	6
octahedron	6	12	8
dodecahedron	20	30	12
isocahedron	12	30	20

There exist only five such polyhedra – the Platonic solids! If we want sharp approximations with, for example, 500 points (vertices) there is no solution with points evenly spaced in this sense, a sad but indisputable fact.

The alert reader may already have asked: why complicate things, why not just choose points corresponding to longitude/latitude coordinates? In other words, describe points on the sphere with a longitude angle φ between 0 and 360°, and a latitude angle ψ between −90° (South pole) and +90° (North pole). Say that we divide the longitude interval (0,360) into n_{long} equal parts and the latitude interval (−90, 90) into n_{lat}, not necessarily evenly spaced parts, we get an arbitrarily large number of points on the sphere. It is true that they are not evenly distributed according to the given definition, this is impossible, but they may give us a useful way of representing the surface.

Of course we are not interested just in the sphere but in other closed surfaces, but the same idea could be applied. For the stomach shape, for example, we could define a North pole at the center of one of its openings and a South pole at the other. Then latitude and longitude circles would give a curvilinear coordinate system on the surface as illustrated in Figure 40.

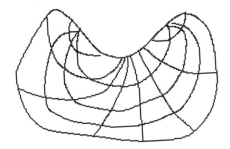

FIGURE 40. LONGITUDE/LATITUDE CURVES ON A STOMACH SHAPE

Such a surface representation could be useful, but it has some drawbacks since it singles out two points, the poles, and treats them differently from the rest. Let us just add that surface representations are needed in many areas of science and technology; one instance is the design of automobile bodies. The designer often expresses the shape by splines with surface patches, where each patch has a simple mathematical shape, given by an algebraic equation for each patch. The patches bordering on each other are joined smoothly so that the derivatives up to some order are continuous across the joint. Expensive cars seem to have more pronounced lines of discontinuities!

So far we have only discussed the surface that bounds the object. If we also want to deal with internal structures the coordinate systems with two "angular" coordinates will have to be augmented with a third "depth" coordinate. Look at the organism in Figure 41: a nematode, *C. elegans*. It consists of 959 cells arranged in a specified order so that a particular cell is known to border certain other cells. This defines certain pairs of cells to be neighbors and this neighborhood relation could be described by a graph (note 3.7). Such a graph can also be thought of as a (discrete) coordinate system.

FIGURE 41. THE 959 CELLS OF C. ELEGANS

In the curvilinear coordinate systems for Figure 40 the φ, ψ coordinates a point with coordinates (φ_i, ψ_j) will have four naturally defined neighbors $(\varphi_{i+1}, \psi_j), (\varphi_{i-1}, \psi_j), (\varphi_i, \psi_{j+1})$, (φ_i, ψ_{j-1}). Well, this is not completely true, for example at the boundary there will be exceptions or at the north and south "poles." Anyway, for the cells in *C. elegans* each cell will be a neighbor by physical contiguity to others in a complicated, not very structured way. But this is the way it is, nature has organized the architecture of *C. elegans* in this way and we have to accept it – the pattern is complex!

This is not enough of difficulties. When the *C. elegans* with its well deserved name moves around to do whatever nematodes do, it changes its form, it is certainly not a rigid body. Therefore it is not enough to describe its architecture in a fixed geometry; we must also represent its variable shape by some mathematical formalism that seems to elude us at the moment. More about this challenging task later.

3.7. Internal patterns. The patterns in the last section had a more complicated dependency structure than the 2-D ones, but this is caused only partly by the fact that they live in 3-D. We shall see now that there are complicated closed patterns also in the plane. We have already touched upon one reason for this: the objects in the patterns are not just set patterns but also have pictorial "events" inside. Look at Figure 42, showing a cell with organelles inside. For the moment it does not matter what the biological significance of the organelles is; we just want to emphasize that such pictures have internal structure. It is not enough to describe them in terms of boundaries.

In Figure 43 we show a schematic view of a cell as it can appear in the gynecological Pap smears. In it we see the nucleus, which has some texture not shown in the figure. The rest of the cell has some other texture. To detect pathological cells a technician observes the textures, the size and shape of the nucleus (regular or not), and other similar properties. Combining these observations a decision will be made whether the cell should be regarded as normal or not.

Let us design a simple information architecture for a Pap smear cell. As a minimum

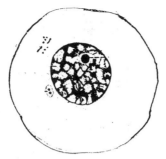

FIGURE 42. CELL WITH ORGANELLES

we have to model the boundary of the cell, the boundary of the nucleus and the textures inside the nucleus, outside the nucleus inside the cell (the cytoplasm), and outside the cell (the inter-cellular matter). In section 3.4 we approximated closed boundaries by polygons, so let us do that for both boundaries in question. Obviously the nucleus boundary is inside the cell boundary so that the two boundaries are subject to some global coupling, but it is not so easy to be more precise than that. An attempt to do this is shown in Figure 43. The elements of the nucleus boundary B_1, the sides in the polygon, are coupled to corresponding elements of the cell boundary B_2. Thus, if B_2 moves in some direction it will try to drag B_1 along in the same direction, so that B_1 will tend to remain inside B_2. The couplings, whatever they really mean, appear as spokes in a wheel with an outer rim B_2 and an inner one B_1.

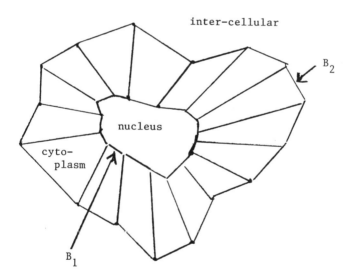

FIGURE 43. SCHEMATIC PAP SMEAR CELL

68 *Chapter 3: Closed Patterns*

Once B_1 and B_2 are given we expect some texture inside B_1, another between B_1 and B_2, and a third outside of B_2. Texture patterns were introduced in section 3.6 and we will leave it at that for the time being.

Look at the attractive design in Plate 5. This pattern is made up of rhombic colored pieces translated and rotated, joined so that they fit together in a symmetric arrangement. The rhombs of a given color are *similar* to each other, meaning that they are all (approximately) congruent to each other in the sense of Euclidean geometry. They are combined along the sides so that one rhomb will neighbor four others. Exceptions to this *rule of combination* occur at the boundary and at the center so that the global architecture is more involved than would appear at first glance. This is typical for patterns with internal structure, whether in 2-D or 3-D.

3.8. Multiple object patterns. Another circumstance that tends to complicate the information architecture is illustrated in Figure 44. It is a high magnification (25,000 x) electron micrograph of a cardiac cell in a mouse. For the moment let us look only at the mitochondria in the picture, the oblong objects with stripes. Arrows point to two of them. The main function of the mitochondria is to produce energy needed in the cell. The mitochondria have a fairly characteristic shape but it varies a good deal and can be vaguely circular or quite elongated. The inside is also very characteristic and different from the outside. Some of the variability is due to the fact that we actually see a 2-D slice of the whole 3-D volume, but the real (biological) variability is the dominating one.

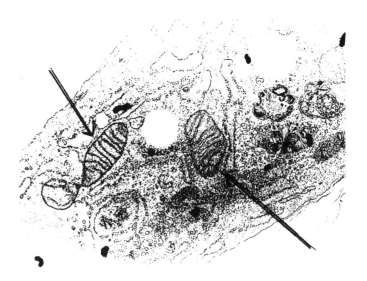

FIGURE 44. MITOCHONDRIA INDICATED BY ARROWS

But this is not what we have in mind just now. In contrast to what we have discussed so far we now deal with an image ensemble where the number of objects varies from one image to another. For a given picture, before it is analyzed, we do not know the *location, orientation,* or *shape* of the mitochondria, and to make things worse for the analyst (human or machine) we do not even know their *number*! How can one handle this difficulty?

Some introspection is in order. When you look at the picture, how do you decide on where the mitochondria (or other objects) are, assuming that you have some vague prior knowledge of how they are supposed to look? Well, their boundaries may be possible to detect since the inside and outside are so different. Sometimes one may even be able to see the membrane that separates inside/outside. It is true that this is not always so; sometimes other matter or optical imprecisions prevent us from seeing a completed, closed, boundary curve. A cursory inspection will lead us to make conjectures about where the mitochondria may be. We form tentative hypotheses and look more closely at these parts of the picture. This may lead us to conclude that what we thought was a mitochondrion is something else. On the other hand, we may discover new places in the picture worth looking at more closely.

By such a *sequential process of hypothesis formation* we are led to conclusions that may or may not be completely correct. Of course this is pure speculation: we do not really know the search strategy that trained cytologists use to find objects of interest and they may not know it exactly themselves. Also, the description is not precise enough, and we can only make the modest claim that the above strategy does not look unreasonable. The American philosopher C. S. Peirce coined the term *abduction*, as a complement to deduction and induction, for the thought process by which we form and modify hypotheses. Therefore the above could be described as the study of abduction algorithms.

But we can aid the formation of hypotheses. Indeed, we know something in advance about the size of the mitochondria. We do not expect them to be densely packed, nor completely absent in the picture. This implies that we have some, admittedly imprecise, knowledge about the number even before we have looked at the picture.

Similarly, the number of cells in Figure 45 is not known in advance, but we can make conjectures. The inspection of the Pap smears is done by trained technicians, each of whom may be handling hundreds of pictures in a single day. The work is monotonous but requires great diligence considering the severe consequences if abnormal cells are not spotted. In some states laws limit the number of smears examined per day; this is to make sure that the technician's alertness does not slacken as time goes by. Much attention has been devoted to the problem of automating the process, wholly or in part. The abnormal variations are so varied that it is no simple matter to code them into behavioral algorithms, just as was the case for the mitochondria micrographs.

How does one represent and handle imprecise knowledge by precise algorithms? This is a fundamental issue that we shall return to many times.

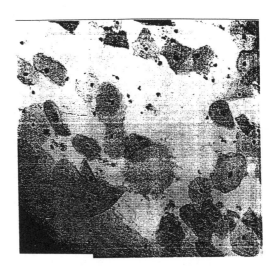

FIGURE 45. PAP SMEAR

3.9. Pattern interference. So far we have examined one pattern at a time, but some fascinating images are the result of the interaction between two patterns. Borrowing a term from optics and the study of other wavelike phenomena we shall then speak of *interference between patterns*.

Returning to the weaving patterns in section 3.2, we saw how periodic weaves could be represented by local coordinates. This observation will help us later but it does not express how weaves are actually produced with warp and weft, and so on. To do this consider again the weaves in the figures of section 3.2.

Just as for weave patterns it happens often in the study of patterns that we have more than one type of representation. We will then have to choose between them, considering their simplicity, degree of approximation and, in particular, how well they express the subject matter knowledge available about them. For weave patterns we shall give an alternative representation in section 8.6 in addition to the analysis in section 3.2: interference between warp and weft patterns.

Another famous type of interference in pictorial patterns is obtained by letting two families of closely spaced curves in the plane interact. Introduce one family F of curves, *curve patterns*, expressed by the equation

$$F(x,y) = k; \quad k = \cdots -2, -1, 0, 1, 2, \cdots$$

where F is some function of the two coordinates x and y in an ordinary (rectangular) coordinate system. For any integer k this defines a curve in the plane. Also introduce a second family G of curves

$$G(x,y) = \ell;\ \ell = \cdots -2, -1, 0, 1, 2, \cdots$$

When we plot these curves we get pictures like the one in Figure 46, with lots of small quadrilaterals. It has been known for a long time (note 3.11) that if the quadrilaterals are long and narrow an observer will get a remarkable illusion of apparent curves. These curves are not really there; they are subjective and can be thought of as the interference between two line patterns. In Figure 46 the interference curves, the *Moiré fringes*, are shown as dotted curves, but in order to see them clearly the quadrilaterals have to be made more extreme in their length/width ratio. We have illustrated this by letting both F and G consist of parabolas with slowly varying parameters. It produces the attractive image in Figure 47.

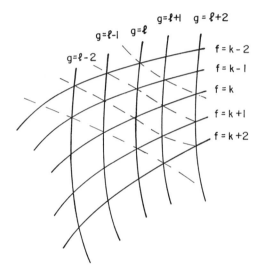

FIGURE 46. MOIRÉ FRINGES

3.10. Pattern of speculation. If there are any patterns in human reasoning they must be, almost by definition, in stereotyped thinking, not in original thought. Are there prototypes of reasoning? As usual it is best to go back to the ancient Greeks, in particular to Aristotle. In his logic an important role is played by *syllogisms*. One example is

all men are mortal

Socrates is a man

therefore Socrates is mortal

FIGURE 47. PARABOLIC MOIRÉ SYSTEM

another is

no dog is human
all poodles are dogs
therefore no poodle is human

In all there are nineteen valid syllogisms according to Middle Ages schoolmen. Abstractly we can represent the first one as

all x are y
z is x
z is y

and the second one as

no x is y
all z are x
no z is y

where one is allowed to substitute various concepts for x, y, z. Anticipating the discussion in section 6.3 we can speak of these abstractions as *templates* in which we can substitute various (logical) elements.

Syllogisms are striking examples of abstract reasoning but their importance should not be exaggerated. Bertrand Russell, who was not only a pioneering philosopher but also a wit, once claimed that the only time he had used syllogisms was when deciding whether to bring an umbrella or not when the weather forecast said rain.

More recent systems of logical reasoning, say, propositional calculus (note 3.12) deal with formulas like (∼ means "not," ∨ means "or," ∧ means "and")

$$\sim (A \vee B) = (\sim A) \wedge (\sim B)$$

as

not (feline or canine) means (not feline) and (not canine)

Here the primitives are simple statements about something (feline) and logical operations like "not," "and," "or" and perhaps others. Note the similarity to formal languages and also how we can start from a template like the abstract expression

$$A \wedge (\sim B)$$

and substitute particular statements to get

(it is raining) and (it is not hot).

We combine primitives into configurations following some (grammatical?) rules.

Perhaps one could make Russell's umbrella decision by a reasoning visualized by the diagram in Figure 48. The boxes represent information, more or less certain, and the arrows mean inferences of different credibility. This diagram has a loop in it: it is *parallel* rather than sequential, and it contains uncertainties; it is *non-deterministic* rather than rigidly logical.

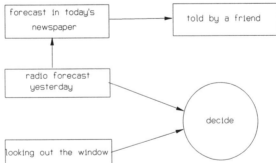

FIGURE 48. REASONING DIAGRAM

The diagram visualizes how we may rank the sources of information that influence the decision. Yesterday's radio forecast probably influences the forecast in today's newspaper, which in turn may have made our friend tell us what he thinks the weather will be, and so on. When we decide we may have all or some of these information sources available. Scientific theories can be looked at from the same perspective. Henri Poincaré, one of the giants in modern mathematics, once described theories as "atoms of thought hooked together." The atoms are simple statements and the glue that binds them together is a logic of some kind.

To be specific let us look at the way we can build up hypotheses for fitting a curve, a function, to data. Say that we are looking at a function $f(x)$ of a single real variable x, $z = f(x)$, and are given a set of x-values with corresponding z-values. This is like the psychological intelligence tests that used to be trusted in the past. For example we are given the table

$$\begin{cases} x = 1\ 2\ 3\ 4\ 5\ 6 \\ z = 4\ 8\ 12\ 16\ 20\ 24 \end{cases}$$

and immediately guess the relation $z = 4x$. Of course this is not necessarily true since the series could continue with $x = 7, z = 35$, but the guess seems reasonable. Why? Well, behind the guess is hidden a conscious or unconscious application of Occam's razor (note 3.13): use the simplest possible explanation of the observed data. But what is "simplest"? Informally, it may mean combinations of the elementary arithmetic operations: add, subtract, multiply, divide, perhaps augmented by other elementary functions like $\sqrt{x}$, $\log x$, and so on.

The table

$$\begin{cases} x = 1\ 2\ 3\ 4 \\ z = 3\ 8\ 15\ 24 \end{cases}$$

can be explained by $z = x^2 + 2x$. The table

$$\begin{cases} x = 1\ 2\ 3\ 4\ 5\ 6\ 7 \\ z = 2\ 3\ 5\ 7\ 11\ 13\ 17 \end{cases}$$

is harder. One explanation is that the z-values are the prime numbers, and early students of number theory used to search for *simple* functions $f(x)$ that would represent the prime numbers; an attempt that failed.

We could just as well study functions of more than one variable: $z = f(x, y)$ has two variables x and y. If we are given the values

$$\begin{cases} x = 1\ 5\ 10\ 16 \\ y = 3\ 2\ 1\ 4 \\ z = 4\ 7\ 11\ 20 \end{cases}$$

one guesses naturally that the function is simply $f(x, y) = x + y$. The table

$$\begin{cases} x = 1\ 5\ 10\ 16 \\ y = 3\ 2\ 1\ 4 \\ z = 9\ 20\ 10\ 256 \end{cases}$$

may take several guesses and could result in the hypothesis $f(x,y) = xy^2$.

Now a more complicated function, $f(x,y) = (x+y)\sqrt{y} - 3x$ visualized in the *flow diagram* in Figure 49. Here the square boxes represent operations: addition, multiplication, and subtraction, each with two inputs, and square roots, with a single input. The round boxes represent assignments, some of which have variable names like x and y, others have numerical specification like "3". The connection architecture is more complex than in earlier cases. These *computational modules* are joined together by arrows that show how data flow through the network. The arrow on the right of the diagram stands for output.

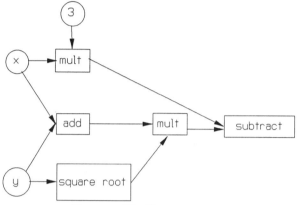

FIGURE 49. FLOW DIAGRAM

We argue that such a diagram represents a *hypothesis* about functional relationship that describes how inputs give rise to outputs. Or one could say that it visualizes a *pattern of computing* (reasoning in numerical terms). Physicists, engineers and others are trying to guess such patterns in their daily work, sometimes with success, and this naturally leads to the question whether it could be done equally well by an algorithm – could a machine (an abduction machine!) be made to recognize, generate and hypothesize, that is, to understand such patterns?

Some readers will recognize such a task as an instance of AI, *artificial (or machine) intelligence*. AI aims at simulating human intelligence by computer programs, but there has been a lot of controversy (note 3.14) about what this means. A celebrated definition from the English logician/mathematician Alan Turing (note 3.15) goes as follows. Say that we direct questions to two "boxes," one containing a computer and the other a person. The communication could be via the keyboards of electric typewriters, and we choose questions in order to figure out from the answers which box contains the computer. If we are not able to conclude in a definitive way which box contains a machine we say that the computer program realizes AI.

It should come as no surprise that AI is a charged issue: statements about the possibility of AI often provoke fierce antagonism, a bit reminiscent of the resistance to Darwinism and evolutionary theory in the second half of the nineteenth century (and even today!). We, or many of us, feel offended by claims that seem to reduce us to machines or animals. Feelings about AI were further aggravated by exaggerated optimism and premature assertions by the early AI proponents about what AI would achieve in the near future. Even a skeptical bystander must admit, however, that remarkable achievements have been made. Among the most publicized are the chess-playing programs that play extremely good if not imaginative chess. More practical are the programs for diagnosis in certain branches of medicine, often based on *expert systems* (note 3.16); such systems attempt to formalize the expertise of physicians and are built by systematic interrogation of the experts, *knowledge engineering*.

The umbrella example of automated reasoning by uncertain parallel logic was just a toy example. In a more serious vein we ask if it is possible to achieve *"pattern understanding"* by algorithmic means. To fix ideas, can we construct algorithms that can analyze some of the medical image ensembles we have encountered, and do it as well as humans?

If this can be done, and we shall study this later on in limited contexts, it is doubtful whether the result would deserve the label artificial intelligence. Indeed, what we are talking about is an extremely thin sector of human intelligence, for example the ability to find the mitochondria in an electron micrograph. AI, on the other hand, strives toward a much more ambitious goal, the simulation of human intelligence in a more general sense.

3.11. Speculation about patterns. Many, if not all, the patterns we have discussed can be mathematically formalized as will be done in Part II. Now as we approach the end of our catalogue of patterns, we shall offer some speculations with a less solid basis. These musings will not be followed up in this volume – perhaps they will stimulate some reader to do so.

We mentioned in section 2.2 that literary style patterns may be quantitatively characterized by probabilities attached to syntactic constructs. Could one not push this one step further and operate on a semantic level? In other words deal with the meaning of the text, not just its grammatical structure? Compare with the discussion in section 2.10 about narratology, the analysis of the telling of stories.

Returning to syntax in languages, much of the enormous body of knowledge amassed during centuries of grammatical study is based on a more or less tacit assumption: language is a logical structure that can be precisely described by a finite number of rules.

One is again reminded of Einstein's famous saying that one of the most remarkable features of the universe is that it can be understood. This probably meant "understood" in a precise mathematical sense and referred to physics. But what is true for physics need not be true for biology, and language is a biological product. There is no reason to believe that language with its unlimited variability could be described accurately by rigid rules, at least not if their number is moderate, say 1000 or so. Already the very notion "grammatically correct" is suspect.

If this view is accepted, and it is not by everyone involved, it seems reasonable to ask whether less rigid "grammars" could be constructed that do not offer a yes/no answer to the questions of grammaticality of sentences. This would be in the hope of finding a formalization of real language as it is actually spoken. Much attention has been given to this topic in recent years (note 3.17).

How would one analyze sentences like the following that normally would not be accepted as grammatical sentences:

> what is the next flight to – uh – LAX?
> I really admire
> the comb and brush

just to mention what the author heard the day of writing this. The reader can easily add more examples of syntactically incorrect but more or less understandable sentences.

It is far from obvious how a theory of colloquial language should be developed. Another doctrine that one could try to formalize is Freudian psychoanalysis. To diagnose a patient's troubles certain basic concepts are combined: the Oedipus complex, the id, censoring, the unconscious, the castration complex and others. These are the elements that form analytic patterns but it is less clear what the exact rules are that tell the analyst how to connect them. Perhaps one should not ask such a question. As Niels Bohr said, it is good to express oneself precisely in science but there is no need to talk more clearly than one's own thinking. There are limits for mathematical formalizations; some bodies of knowledge are so vague that mathematization would add only spurious rigor.

However that may be, our catalogue of patterns has given us hints about what principles we should try to build on:

a. *Generate* patterns from basic constructs, elementary concepts, primitives.

b. *Connect* the elements by designing rules of combination allowing for variability.

c. *Interpret* the resulting configurations, what do they really mean.

d. *Relate* these abstractions (combinations of elements) to what can be "seen" by real persons/sensors with all their imperfections.

We shall now build a mathematical *formalism of patterns* and then apply it to several of the patterns in the catalogue.

PART II

THEORY OF PATTERNS

Chapter 4

Analyzing Patterns

In order to understand patterns and analyze their structure we shall formalize the ideas we have encountered in Part I in terms of mathematical constructs. This will enable us to be more precise.

To do this we need a mathematical pattern formalism: *a pattern algebra*. The price we pay for the increase in logical rigor is that the formalism will appear, at least to begin with, somewhat abstract, but we shall try to clarify the ideas behind the constructs by an example that will accompany the formal developments throughout this chapter. In Chapters 5 and 6 we will meet a large number of other examples of pattern types that illustrate the intuitive meaning of these constructs.

4.1. Generators. We shall build representations of patterns from simple building blocks that will be referred to as *generators* and denoted by $g, g_1, g_2, g_i, g_i^0 \ldots$ with subscripts/superscripts as needed. The set of generators needed for a particular knowledge representation, the *generator space* G, will vary from case to case but at the moment we shall let it be quite general.

To represent *symmetries* and *invariances* of patterns we shall use some group of transformation of G onto G (note 4.1), the *similarity group* S with elements denoted by s, $s \in S$. Each s shall mean a bijective mapping $s : G \longleftrightarrow G$, and we shall speak of s as a similarity. We shall assume that the generator index α, see below, is S-invariant so that g and sg are in the same set G^α, $\forall s \in S$.

Sometimes it is natural to split up the whole generator space G into parts G^α, where α

is called the *generator index* varying over some space A. We shall then have a partition

$$G = \bigcup_{\alpha \in A} G^\alpha$$

into disjoint subsets G^α.

To be able to combine such building blocks, the g's, into larger structures each generator g will possess bonds $b_1, b_2, \ldots b_\omega$ where $\omega = \omega(g)$, the *arity* of g, means the number of bonds and can vary from generator to generator. To each bond b_j corresponds a *bond value* β_j from some *bond value space* B.

Bonds may carry markers, for example labels "in" or "out," to indicate that the bonds are directed in, toward g, or out, away from g. For any $g \in G$ the notation $B_s(g)$ will mean the set $\{b_j;\ j = 1, 2, \ldots \omega(g)\}$ where b_j means the *bond coordinate j* together with the markers in the bond, if any, and $B_v(g)$ will mean the set $\{\beta_j;\ 1, 2, \ldots \omega(g)\}$. We shall denote by $B(g)$ the combination of bond structure $B_s(g)$ and bond values $B_v(g)$.

The set $B_s(g)$, the *bond structure* is assumed to be S-invariant, that is if g_1 and g_2 are *similar*, meaning that there exists a similarity s such that $g_2 = sg_1$, then g_1 and g_2 should have the same bond structure. Usually they will have different *bond value sets* B_v however.

It will sometimes be natural to construct the generator space G from a smaller set G_0, the *initial generator space*, $G_0 \subset G$. There will then be a group BSG, usually distinct from the similarity group S, and this *bond structure group BSG* will extend G_0 to G so that any $g \in G$ is obtained (perhaps in more than one way) from some $g_0 \in G_0$, $g = \mu g_0$, $\mu \in BSG$.

We illustrate these concepts by the following

Example. Consider generators given diagrammatically by Figure 1.

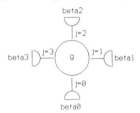

FIGURE 1. GENERATOR DIAGRAM

Here g has arity $\omega = \omega(g) = 4$, $B_s = \{0, 1, 2, 3\}$ and $B_v = \{\beta_0, \beta_1, \beta_2, \beta_3\}$. To each g is associated a pair of integers $(x, y); x, y \in \mathbb{Z}$ which will later be interpreted as coordinates in the plane. The bond structure group BSG will consist of rotations of g by $0^0, 90^0, 180^0, 270^0$, so that $BSG = \mathbb{Z}_4$ (note 4.1).

Let the initial generator space G_0 consist of the "boundary generators" in Figure 2 that will be used later on when their meaning will become apparent. The bond value space here is $B = \{0, 1, 2, 3, 4, 5\}$. When we apply BSG to each $g \in G_0$ we get some duplication, resulting in 14 distinct generators, $|G| = 14$.

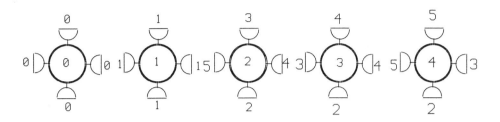

FIGURE 2. AN INITIAL GENERATOR SPACE

We shall let the similarities s be translations in the (x, y)-plane

$$\begin{cases} s : (x, y) \to (x + h, y + k) \\ s = (h, k) \in \mathbb{Z}_2 \end{cases}$$

without affecting $B_s(g)$ or $B_v(g)$. The intuitive meaning of these "boundary generators" is that $g = 0$ represents an inside point, $g = 1$ an outside point, $g = 2, 3, 4$ represent boundary elements that are straight, turning left and right respectively. We shall later use this G for describing patterns with certain *geometric tendencies*.

4.2. Configurations. We can now glue generators together and the bonds will tell us what combinations will hold together. This is a bit like chemistry: atoms (generators) are connected together into molecules (configurations) and the nature of the chemical bonds, ionic, covalent, and so on, decides what combinations of atoms will be stable enough to form molecules.

Say that we try to form a configuration from the generators $g_1, g_2, \ldots g_i \ldots g_n$ where the subscript i will be referred to as a *generator coordinate*. For g_i let the bonds be denoted $b_{i1}, b_{i2}, \ldots b_{ij} \ldots b_{i\omega}$, $\omega = \omega(g_i)$; j is called the *bond coordinates* for g_i. Figure 3 illustrates this for $n - 2$, where the bond (i_1, j_1), with the value β_1, tries to connect with (i_2, j_2), with the value β_2.

84   Chapter 4: Analyzing Patterns

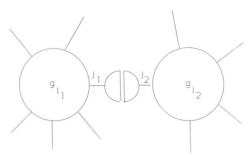

FIGURE 3. INTERACTING GENERATORS

Introduce a *configuration architecture* by selecting a graph, for example the one in Figure 4a where the sites are enumerated by $i = 1, 2, \ldots n;\ n = 6$. At site i we position a generator $g_i \in G$ and then connect the bonds, for example as in the diagram Figure 4b. The resulting structure is called a *configuration*.

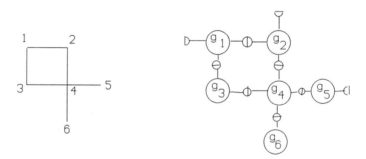

FIGURE 4A. CONNECTOR GRAPH   FIGURE 4B. CONFIGURATION DIAGRAM

We use a graph σ, called a *connector*, that connects some bonds $k_1 = (i_1, j_1)$ with others $k_2 = (i_2, j_2)$, where $k = (i, j)$ labels bonds in general. We shall symbolically represent the configuration as

$$c = \sigma(g_1, g_2, \cdots g_n).$$

Some bonds in c connect to other bonds; they will be called *internal bonds*. The remaining, closed bonds are the *external* ones, the set of them denoted ext(c). In the figure $|ext(c)| = 3$.

We shall often restrict the configuration by local as well as global constraints. On the product space $\mathcal{B} \times \mathcal{B}$ of the bond value space $\mathcal{B}$ crossed with itself we assume, given a truth valued function

$$\rho : B \times B \to \{\text{TRUE,FALSE}\},$$

so that for a pair (β', β'') of bond values the pair is either regular, if $\rho(\beta', \beta'')$ = TRUE or irregular, if $\rho(\beta', \beta'')$ = FALSE. This partitions $B \times B$ into two subsets and ρ therefore is equivalent to a relation, to be called the *bond value relation*. It is not completely arbitrary, we shall assume that it is S-invariant in the sense that if for two internal bonds (i', j') of $g_{i'}$ and (i'', j'') of $g_{i''}$ we have $\rho[\beta_{j'}(g_{i'}), \beta_{j''}(g_{i''})]$ = TRUE then $\rho[\beta_{j'}(sg_{i'}), \beta_{j''}(sg_{i''})]$ = TRUE for all similarities $s \in S$.

At site i the generator g_i sends out bond values $\beta_1(g_i), \beta_2(g_i) \ldots \beta_\omega(g_i)$. To some or all of these bonds b_j there is a bond value, call it β'_j, that comes from the neighboring sites of i in σ and connects to b_j. We shall sometimes use the concept *environment* for the set

$$env(g_i) = (\beta'_1, \beta'_2, \ldots \beta'_\omega).$$

We now introduce the crucial

Definition. A configuration $c = \sigma(g_1, g_2, \ldots g_n)$ is said to be *locally regular* if for any internal bond couple $(i', j') - (i'', j'')$ we have

$$\rho[\beta_{j'}(g_{i'}), \beta_{j''}(g_{i''})] = \text{TRUE}.$$

Local regularity can be formalized in a way that will serve as a guide when we shall relax the regularity in section 4.5. Writing $k = (i, j)$, $k' = (i', j')$ to enumerate bonds in c let us introduce the *first structure formula* as

$$\bigwedge_{<k,k'>} \rho[\beta_j(g_i), \beta_{j'}(g_{i'})] = \text{TRUE}.$$

In this logical conjunction (note 3.11) the notation $<k, k'>$ means that the conjunction is taken over all closed (internal) bond couples (k, k'). The structure formula is just another way of writing Definition 1, but one that will help our intuition later.

We shall also assume that $\sigma \in \Sigma$, some family of graphs called the *connection type*. Among connection types to be used later we mention Σ = LINEAR (all linear chain graphs), Σ = TREE (tree shaped graphs), Σ = LATTICE (graphs with square lattice structure), but many others will also appear.

Definition. A configuration $c = \sigma(g_1, g_2, \ldots g_n)$ is called *globally regular* if $\sigma \in \Sigma$.

If c is both locally and globally regular it is said to be *regular* and the set of all such configurations, the *configuration space* $\mathcal{C}(\mathcal{R})$, where $\mathcal{R} = <G, S, \rho, \Sigma>$ is referred to as a *regularity*.

In Figure 5 consider the subconfiguration c' inside the dotted line; is it regular if the whole configuration $c \in \mathcal{C}(\mathcal{R})$? Since all its internal bond couples remain regular (satisfy the bond relation ρ) it is certainly locally regular. Whether it is globally regular depends upon whether its connector σ' belongs to the connection type Σ as σ does. An important case when this occurs is when the connection type is *monotonic*, which means that if some graph $\sigma \in \Sigma$ then any subgraph σ' of σ also belongs to Σ.

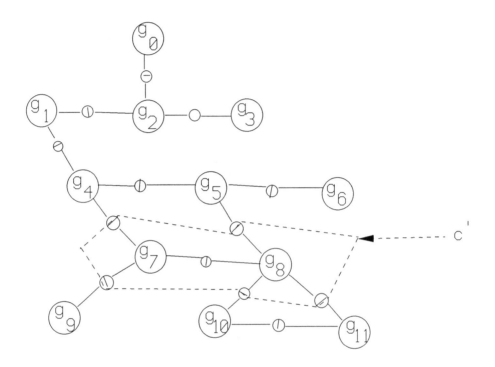

FIGURE 5. CONFIGURATION OF SUBCONFIGURATION

In Figure 5 we can write, with notation to be used in section 4.2,
$$\begin{cases} c &= \sigma(g_0, g_1, g_2, g_3, g_4, g_5, g_6, g_7, g_8, g_9, g_{10}, g_{11}) \\ c' &= \sigma'(g_8, g_9) \\ c'' &= \sigma''(g_0, g_1, g_2, g_3, g_4, g_5, g_6, g_7, g_{10}, g_{11}) = \text{the rest of } c \text{ when } c' \text{ has been removed} \\ c''' &= \sigma'''(g_4, g_5, g_7, g_{10}, g_{11}) = \text{"the external boundary" of } c' \text{ in } c. \end{cases}$$
In other words we can combine regular configuration, for example c' and c'', into larger configurations, say c, not always regular. It is to emphasize this *combinatory* nature

of pattern theory that we have selected the notation $c = \sigma(g_1, \ldots g_n)$ which should be interpreted as meaning that c is a function of $g_1, \ldots g_n$, the variables, and just as in function composition a g_i can be replaced by a configuration c_i if $\text{ext}(c_i) = B(g_i)$.

In Figure 6 we illustrate such a combination of three configurations c', c'', c''' into a fourth one c. Think of the *coupling connector* $\sigma_0(c', c'', c''')$ as a function of three variables c', c'', c''' and taking values in $\mathcal{C}(\mathcal{R})$

$$\sigma_0 : \mathcal{C}(\mathcal{R}) \times \mathcal{C}(\mathcal{R}) \times \mathcal{C}(\mathcal{R}) \to \mathcal{C}(\mathcal{R}).$$

Note, however, that the value in $\mathcal{C}(\mathcal{R})$ of this function is not defined for all c', c'', c''': it may be that the three configurations do not fit, they cannot be glued together. Therefore σ_0 is only a *partial function*, not defined everywhere in $\mathcal{C}(\mathcal{R}) \times \mathcal{C}(\mathcal{R}) \times \mathcal{C}(\mathcal{R})$.

Another type of function, mapping $\mathcal{C}(\mathcal{R})$ onto itself, is given by the similarities applied to $c : c \to sc$; one such function for each $s \in S$. These functions are entire, they are defined everywhere. Compare this to the classical algebraic structures, for example the real line $I\!R$ with the operations $+$ and $\times$. For $x, y \in I\!R$ the operations $x + y$ and $x \times y$ are always defined. We can describe these functions by two tables: one for addition and another for multiplication. The function $\div$ on the other hand is partial, since $x \div y$ is not defined for $y = 0$.

A coupling connector, say with two variables $\sigma(\cdot, \cdot)$, can be described by a *composition table* with two margins, whose values may include $u = $ undefined. Such a table tells us what regular configuration $c = \sigma(c_1, c_2)$ is obtained by coupling c_1 with c_2 via the connector σ, if it is defined (regular) in $\mathcal{C}(\mathcal{R})$. Similarly for coupling connectors with three variables $\sigma(\cdot, \cdot, \cdot)$, and so on, and one table for each similarity in S. The latter table tells us what configuration c' we get by applying a similarity to $c = \sigma(g_1, g_2, \ldots g_n)$, $c' = \sigma(sg_1, sg_2, \ldots sg_n)$; it is automatically defined (regular) since the bond relation ρ is S-invariant and the connector $\sigma \in \Sigma$ is unchanged when we apply s.

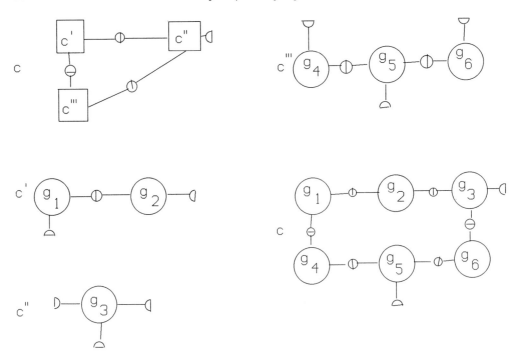

FIGURE 6. COMBINATIONS OF CONFIGURATIONS

Let us illustrate the concepts introduced in this section by our previous

Example. Let ρ be defined by the 6×6 truth valued matrix of bond values β, β' where 1 stands for TRUE and 0 for FALSE. For example, $\beta = 2, \beta' = 3$ fit but $\beta = 0, \beta' = 1$ do not. Introduce the connection type SQUARE LATTICE with all graphs as subgraphs of $\mathbb{Z}^2$, the set of points (x, y) in the plane with x and y integers, and where a site (x, y) is connected to four neighbors $(x+1, y), (x, y+1), (x-1, y), (x, y-1)$. Let S be the translation group $\mathbb{Z}_2$ of translations $(x, y) \to (x + h, y + k)$ with h and k arbitrary integers (note 4.2).

Chapter 4: Analyzing Patterns

		\multicolumn{6}{c}{β}					
		0	1	2	3	4	5
	0	1	0	0	0	1	0
	1	0	1	0	0	0	1
β'	2	0	0	0	1	0	0
	3	0	0	1	0	0	0
	4	1	0	0	0	0	0
	5	0	1	0	0	0	0

A regular configuration can then be shown by a diagram consisting of certain sites of $\sigma \in$ SQUARE LATTICE carrying g-values, here from $G = \{1, 2, 3, 4, 5, 6, 7, 8, 9, 10, 11, 12, 13, 14\}$, and bond couples from $B = \{0, 1, 2, 3, 4, 5\}$, each couple satisfying ρ. The diagram could define a picture like the one in Figure 7 where the arrows indicate generators of the type μg_i, obtained from 2,3, or 4 in Figure 2 of section 4.1, and μ is an element $\mu \in BSG$ (the bond structure group); here BSG means $\mathbb{Z}_4$ consisting of rotations by a multiple of $90°$.

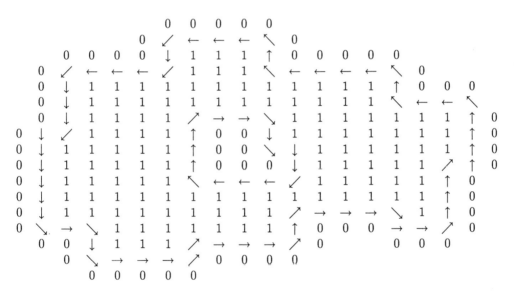

FIGURE 7. CONFIGURATION WITH BOUNDARY GENERATORS

If this configuration is denoted c then sc means the same configuration *shifted* a certain amount up or down and right or left. In addition to this shift-operation in $\mathcal{C}(\mathcal{R})$ we shall sometimes also use a *jittering* operation meaning the transformation

$$\sigma(g_1, g_2, \ldots g_n) \to \sigma(s_1 g_1, s_2 g_2, \ldots s_n g_n)$$

for some vector $(s_1, s_2, \ldots s_n) \in S^n$. This important concept will also appear in section 6.3 as deformable templates. It must be pointed out, however, that the result on the right hand side is not always in $\mathcal{C}(\mathcal{R})$ since regularity may be lost when jittering a regular c. Hence we have only a partial operation: it is not defined for all s-values.

But we shall also have reason to consider mappings of configurations $c \in \mathcal{C}(\mathcal{R})$ to $c' \in \mathcal{C}(\mathcal{R}')$: a different configuration space from the first, but where the mapping preserves structure in some essential way. In analogy with algebra we shall introduce *configuration homomorphisms*.

Definition. Consider two regularities $\mathcal{R} = <G, S, \Sigma, \rho>$ and $\mathcal{R}' = <G', S, \Sigma, \rho'>$ over the same similarity group S and connection type Σ. A mapping $h: \mathcal{C}(\mathcal{R}) \to \mathcal{C}(\mathcal{R}')$ is said to be a homomorphism if h taking $c = \sigma(g_1, g_2, \ldots g_n)$ into $\sigma(g'_1, g'_2, \ldots)$ has the properties

$$\begin{cases} h[\sigma_0(c_1, c_2)] = \sigma_0(hc_1, hc_2) \\ h(sc) = sh(c) \end{cases}$$

where all the appearing configurations are assumed regular.

Remark 1. To simplify we have stated the definition less generally than needed: one can actually allow $\mathcal{R}$ and $\mathcal{R}'$ to have different S and Σ. The role of homomorphisms will become clearer when we illustrate it by examples in the following chapters.

Remark 2. Sometimes it is necessary to restrict the s-values in the definition to a subgroup of S, for example for deformable templates used below (note 4.3).

In some configuration space $\mathcal{C}(\mathcal{R})$ with monotonic connection type Σ consider a particular configuration $c^0 = \sigma(g_1^0, g_2^0, \ldots g_n^0)$; it will be spoken of as a *template* (sometimes we have to consider several templates, but not here). Form a new space of configurations by fixing similarities $s_1, s_2, \ldots s_n$ arbitrarily except that we demand that the "jittered" configuration $\sigma(s_1 g_1^0, s_2 g_2^0, \ldots s_n g_n^0)$ be regular, belong to $\mathcal{C}(\mathcal{R})$, which is not automatically true. Consider the set of configurations of the form $c' = \sigma'(g_{i_1}^0, g_{i_2}^0, \ldots g_{i_n}^0)$ where σ' is a subgraph of σ with sites $i_1, i_2, \ldots i_n$. On this space $\mathcal{C}_0$, the set of *partial templates*, we allow combinations of configurations if the result still belongs to $\mathcal{C}_0$.

The partial template spaces play an important role in knowledge representations of variability. For any $c' \in \mathcal{C}_0$ we define $c'' = h(c') = \sigma'(s_{i_1} g_{i_1}^0, s_{i_2} g_{i_2}, \ldots s_{i_m} g_{i_m}^0)$ which is automatically regular, $c'' \in \mathcal{C}_0$. This mapping h satisfies the first condition in the definition of homomorphisms since if $c'_1, c'_2, c' = \sigma''(c'_1, c'_2)$ are all regular then $h[\sigma''(c'_1, c'_2)] =$

$\sigma''(hc'_1, hc'_2)$. The second condition demands that, for any $s \in S$, we have with the same notation

$$h(sc') = h[\sigma'(sg^0_{i_1}, sg^0_{i_2}, \ldots sg^0_{i_m})] =$$
$$= \sigma'(s_{i_1}sg^0_{i_1}, s_{i_2}sg^0_{i_2}, \ldots s_{i_m}sg^0_{i_m}) =$$
$$= sh(c') = s\sigma'(s_{i_1}g^0_{i_1}, s_{i_2}g^0_{i_2}, \ldots s_{i_m}g^0_{i_m}) =$$
$$= \sigma'(ss_{i_1}g^0_{i_1}, ss_{i_2}, \ldots ss_{i_m}g^0_{i_m})$$

but only if all $ss_i = s_i s$. This is the case if S is commutative but can be interpreted under more general conditions (see note 4.3). We shall speak of such homomorphisms as producing *deformed templates*. Sometimes we shall let the s_i be randomly chosen and we then refer to *probabilistically deformed templates*; many examples of this will appear later and make the concept more intuitive.

4.3. Images. The configuration space is the first of the *regular structures* that we are building. The next one consists of *images*, a concept that formalizes the idea of *observables*. In other words, a configuration is a mathematical abstraction which typically cannot be observed directly, but the image can. An ideal observer, with perfect instrumentation so that the sensor used has no observational errors, will be able to see some object, called image, that may carry less information than the configuration that is being observed. The loss of information is not caused by noise in the sensors (that will be treated in section 4.6) but is more fundamental and it takes some care to formalize the concept of image in order to get a suitable algebraic structure.

To do this we introduce an equivalence relation (note 4.4) R between elements in $\mathcal{C}(\mathcal{R})$, so that for $c_1, c_2 \in \mathcal{C}(\mathcal{R})$ they are either equivalent, $c_1 \equiv c_2(mod R)$ or they belong to different equivalence classes in the partition $\mathcal{I} = \mathcal{C}(\mathcal{R})/R$. We ask that R has the following properties that will later on be seen to be needed. R will be called an identification rule. We shall write $I = [c]_R$ for the image that contains c.

Definition. An equivalence relation R in $\mathcal{C}(\mathcal{R})$ is said to be an identification rule

$$\begin{cases} (i) & \text{if } c \equiv c'(mod R) \text{ then } ext(c) = ext(c') \\ (ii) & \text{if } c \equiv c'(mod R) \text{ then } sc \equiv sc'(mod R), \forall s \in S \\ (iii) & \text{if } c_1, c_2, c'_1, c'_2 \text{ as well as } c = \sigma(c_1, c_2), c' = \sigma(c'_1, c'_2) \end{cases}$$

are all regular and if $c_i \equiv c'_i(mod R)$; $i = 1, 2$; then $c \equiv c'(mod R)$.

The equivalence classes in $\mathcal{C}(\mathcal{R})$ are then called *images* and the set of images in the quotient space

$$\mathcal{I} = \mathcal{C}(\mathcal{R})/R$$

is called an *image algebra*. The images can be treated in a combinatory way as shown by

Theorem. *The similarity transformations can be uniquely extended to $\mathcal{I}$ and connectors σ to part of $\mathcal{I} \times \mathcal{I}$ in such a way that*

$$\begin{cases} (i) & s_1(s_2 I) = (s_1 s_2) I; \; I \in \mathcal{I} \\ (ii) & s\sigma(I_1, I_2) = \sigma(sI_1, sI_2) \; if \; \sigma(I_1, I_2) \in \mathcal{I}. \end{cases}$$

Proof: The natural extension of $s : \mathcal{C}(\mathcal{R}) \to \mathcal{C}(\mathcal{R})$ to $\mathcal{I}$ is to define $sI = [sc]_R$ where $c \in I$. To see that this definition is unique let c' be another configuration in $I \subset \mathcal{C}(\mathcal{R})$. But condition (ii) in the definition tells us that $[sc']_R = [sc]_R$ and this proves uniqueness.

Similarly for given I_1 and I_2 in the image algebra $\mathcal{I}$ consider two regular configurations $c_1 \in I$ and $c_2 \in I$. If $\sigma(c_1, c_2)$ is meaningful as a regular configuration define $\sigma(I_1, I_2) = [\sigma(c_1, c_2)]_R$. Uniqueness of this definition follows from condition (iii) in the definition.

To see that (i) holds in the statement of the theorem we observe that, if $c \in I$,

$$s_1(s_2 I) = s_1(s_2 [c]_R) = s_1[s_2 c]_R = [s_1 s_2 c] = (s_1 s_2) I$$

using condition (ii) in the definition. Also, to prove statement (ii), if $c_1 \in I_1$ and $c_2 \in I_2$,

$$s\sigma(I_1, I_2) = s[\sigma(c_1, c_2)]_R = [s\sigma(c_1, c_2)]_R = [\sigma(sc_1, sc_2)]_R \\ = s\sigma(I_1, I_2).$$

Q.E.D.

Hence we can deal with the I's as an element of $\mathcal{I}$ and form composition tables in $\mathcal{I}$ as earlier in $\mathcal{C}(\mathcal{R})$, so that $\mathcal{I}$ forms an algebra. At the moment we shall not study its properties any further but go back to our example to illustrate the idea of images.

Example. The configuration in Figure 6 consists of generators that represent inside points, from $g = 1$ in the initial generator space G_0, and outside points corresponding to $g = 0$, and boundary points that correspond to $g = 2, 3, 4$. Define R by saying that c' and c'' are R-equivalent if at each site i, g'_i in c' and g''_i in c'' belong to the same of these three types and if the external bonds of c' equal the external bonds of c''. It is then seen that this R satisfies the conditions in Definition 1, it is a legitimate identification rule and defines an image algebra.

Such images can be represented schematically as in Figure 8; compare with Figure 7. The hatched area means the inside points; boundary points correspond to the horizontal and vertical lines shown, and the rest are outside points. The diagram is not quite complete; it ought to also include the external bonds.

A similarity $s \in S = \mathbb{Z}^2$ applied to I will just shift its location in a rigid manner over the square lattice. Two images I_1 and I_2 can be combined if they fit via $\mathcal{R}$; this means simply combining two pictures assuming that this is geometrically meaningful.

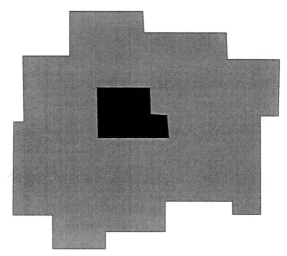

FIGURE 8. A BOUNDARY IMAGE

4.4. Patterns. The word "pattern" in everyday parlance often means something typical that can be repeated, something characteristic capable of repetition into similar copies. The textile patterns in section 3.2, for example, clearly have this flavor. The operative term here is "similar," which we have given a precise meaning in terms of a similarity group S.

This leads us naturally to the following formalization of "pattern."

Definition. Given an image algebra $\mathcal{I}$ we shall understand by a pattern any S-invariant subset $\mathcal{P}$ of $\mathcal{I}$, and by a pattern family $\{\mathcal{P}_\gamma\}$ a partition of $\mathcal{I}$ into patterns.

In other words a subset $\mathcal{P} \subset \mathcal{I}$ is a pattern if for any $s \in S$ and $I \in \mathcal{P}$ we have also $sI \in \mathcal{P}$. We shall be particularly interested in *minimal patterns* that have no pattern as a proper subset. This concept leads to the simple

Proposition. The finest pattern partition $\{\mathcal{P}^p\}$ of $\mathcal{I}$ is uniquely determined and can be generated by templates $T^p \in \mathcal{P}^p$ such that $\mathcal{P}^p = ST^p, \forall p$.

Proof: Let us define a relation on $\mathcal{I}$ by saying that I_1 and I_2 are similar, formally written as $I_1 \equiv I_2 (mod S)$ if there exists an $s \in S$ such that $sI_1 = I_2$. It is obvious that this relation is reflexive, $I \equiv I$, symmetric, $I_1 \equiv I_2 \implies I_2 \equiv I_1$, and transitive, $I_1 \equiv I_2$ and $I_2 \equiv I_3 \longrightarrow I_1 \equiv I_3$. Hence it is an equivalence and induces a partition of $\mathcal{I}$ into equivalence classes.

From each such equivalence class select one representative, a *template*, and denote it T^p with some suitable superscript p. This choice is, in general, quite arbitrary, but in particular situations there is often a natural choice of the representation T^p.

Setting $\mathcal{P}^p = ST^p$ is clear that $\mathcal{P}^{p_1}$ and $\mathcal{P}^{p_2}$ are disjoint if $p_1 \neq p_2$. Indeed, if they had one element in common, say I, we would have $I = s_1 T^{p_1} = s_2 T^{p_2} \Longrightarrow T^{p_1} = s_1^{-1} s_2 T^{p_2} \Longrightarrow T^{p_1} \equiv T^{p_2}$ against our construction. It is also obvious that $\bigcup_p \mathcal{P}^p = \mathcal{I}$ and that each $\mathcal{P}^p$ is S-invariant so that $\{P^p\}$ is a partition pattern family.

But $\{P^p\}$ is also minimal, since otherwise some pattern, say $\mathcal{P}^p$, must have an S-invariant proper subset P_0^p. The latter is impossible since then, if $I_0 = sT^p \in P_0^p$,

$$P_0^p = SP_0^p = SI_0 = SsT^p = ST^p = \mathcal{P}^p.$$

This shows the existence of a minimal pattern partition.

Uniqueness is immediate, since if $\{\mathcal{P}^p\}$ is an arbitrary invariant pattern partition and if $I \in \mathcal{P}^p$ then $SI \subseteq S\mathcal{P}^p = \mathcal{P}^p$. But SI cannot be a proper subset of $\mathcal{P}^p$ which implies uniqueness.

Q.E.D.

Example: Consider the picture in Figure 1 and form the set $\mathcal{P}$ of all possible translations of it. Then the set $\mathcal{P}$ is clearly invariant with respect to $S = \mathbb{Z}^2$ and it is also seen to be minimal. This concept of a pattern is therefore a simple and attractive one. When we build our mathematical knowledge representations for a particular application the crucial and difficult task is in the construction of realistic configuration spaces and image algebras. This done the third regular structure – pattern families – is then directly obtained. It should be mentioned, however, that it is sometimes necessary to employ more than one similarity group.

4.5. Probabilities. The classical algebraic structures support probability distributions on the real line $\mathbb{R}$, the integers $\mathbb{Z}$, and so on. Indeed, much of probability theory studies what happens with such probability distributions when we combine them. For example, on $\mathbb{R}$ we are led to the concept of a real valued random variable. If we have two distributions P_1 and P_2 on $\mathbb{R}$ and consider two independent random variables x_1 and x_2 with these distributions respectively and let x_1 and x_2 be stochastically independent, then the sum $x_1 + x_2$ has as its probability distribution the convolution $P_1 * P_2$. Repeating this for several random variables probabilists are accustomed to study what happens to repeated convolutions. This led to the law of large numbers, the central limit theorem and a multitude of other important results in probability theory. It is natural to ask if we can build a similar probabilistic structure on top of the algebraic one of $\mathcal{C}(\mathcal{R})$ and $\mathcal{I}$ in such a way that this superstructure is a natural extension of the algebraic one.

The concept of configuration space is built on the combinatory idea: we combine generators that "fit" each other, which is decided by evaluating the bond relation ρ on $B \times B$. But ρ is binary, it can only take the values TRUE or FALSE, and it is natural to extend this to some continuum valued function $A(\cdot,\cdot)$ on $B \times B$. To begin with we shall only assume that A, the *acceptor function*, takes non-negative real values. A hint for doing this is given by the first structure formula in section 4.2.

To simplify let us first deal with the case where G is a finite set and σ is fixed. We shall associate probabilities to the configuration c by a *second structure formula* (compare with the notation in section 4.2)

(1) $$p(c) = \frac{1}{Z} \prod_{<k,k'>} A[\beta_j(g_i), \beta_{j'}(g_{i'})]$$

where Z is a constant that normalizes the probabilities so that

$$\sum_c p(c) = 1.$$

An alternative expression that will sometimes be used is

(2) $$p(c) = \frac{1}{Z} \prod_{<k,k'>} A[\beta_j(g_i), \beta_j(g_{i'})] \prod_i Q(g_i)$$

where $Q(\cdot)$ is a non-negative weight function defined on G. Its role is to make probabilities depend not only upon couplings $\beta - \beta'$ but also on the generators g themselves.

A reader familiar with statistical mechanics will recognize (1) as a Gibb's measure (note 4.5) and Z as the *partition* function. There exponential notation is often used

$$A(\beta, \beta') = \exp[-\frac{1}{T} E(\beta, \beta')]$$

where E is called the interaction *energy* and T is a positive constant, the *temperature*. Sometimes we shall write instead

$$A(\beta, \beta') = \exp[a(\frac{\beta - \beta'}{\epsilon})]$$

which is natural when the bond value space B forms a vector space, so that the difference $\beta - \beta'$ is meaningful. The *coupling constant* ϵ plays a similar but different role to that of temperature, and the function $a(\cdot)$ is known as the *affinity*. The total energy and affinity then mean simply

$$\begin{cases} E(c) = \sum_{<k,k'>} E(\beta_k, \beta_{k'}) \\ a(c) = \sum_{<k,k'>} a(\beta_k, \beta_{k'}). \end{cases}$$

We shall denote this *relaxed regularity* by $\mathcal{R} = <G, S, A, \Sigma>$ and will start with a rigid regularity $\mathcal{R}_{\text{rigid}} = <G, S, \rho, \Sigma>$ and then replace the truth valued bond value function ρ by the numeric valued acceptor function A. This will be needed when we try to create knowledge representations of pattern variability. This probability superstructure on top of the algebraic one in $\mathcal{C}(\mathcal{R})$ automatically induces probabilities in the image algebra by the natural relation for $E \subset \mathcal{I}$

$$P(E) = P(\text{set of all } c \text{ such that } [c]_R \in E).$$

Similarly, probabilities are induced in a pattern family since the patterns are just subsets of $\mathcal{I}$; we shall then speak of *metric patterns*.

When $T \downarrow 0$ or $\epsilon \downarrow 0$ the probabilities will concentrate on the set of configurations/images for which the energy $E(c)$ is close to minimum or affinity close to maximum. Such an image ensemble is said to be *cold*, and if the limit exists for $T \downarrow 0$, $\epsilon \downarrow 0$ we speak of *frozen images*. They tend to have a rigid regularity $<G, S, \rho, \Sigma>$ with a ρ related to limits of $A(\cdot, \cdot), a(\cdot)$. On the other hand if T or ϵ are large we speak of *hot images*; they are more random in character.

If G forms a continuum the left side $p(c)$ in the second structure formula will be interpreted as a probability density with the obvious consequences that this entails. More important is the following observation that will be made at first only for a fixed connector $\sigma \in \Sigma$. Consider Figure 2 in section 4.1. A probability distribution over the configuration space makes the generator g_i random; they are random variables taking values in G and $p(c)$ describes their joint probability distribution. Let us now consider the random subconfiguration $c' = \sigma'(g_8, g_9)$ and ask what is the conditional distribution of c' given the remainder c'' with the notation from section 4.2.

We have
$$c = \sigma_0(c', c'').$$

Let us calculate the conditional probability

$$P(c'|c'') = \frac{P(c' \text{ and } c'')}{P(c'')} = \frac{P(c)}{P(c'')}.$$

The partition function Z cancels since it appears in both numerator and denominator and

$$P(c'|c'') = \left(\prod_{(k,k') \in \sigma} A[\beta_j(g_i), \beta_{j'}(g_{i'})] \right) \Big/ \sum \left(\prod_{(k,k') \in \sigma''} A[\beta_j(g_i), \beta_{j'}(g_{i'})] \right)$$

with the notation $k = (i, j), k' = (i', j')$ for the generator/bond coordinates, and where the summation extends over all generators in c'.

But many factors occur both in the numerator and denominator and can be canceled. Indeed, all generators that are not connected with the subconfiguration c' cancel, so that we get with obvious shorthand notation

$$P(c'|c'') = \frac{\prod_{(k,k')\in\sigma'} A}{\prod_{(k,k')\in\sigma'''} A}$$

where the new subconfiguration c''' (in the example $\sigma'''(g_4, g_5, \ldots)$) consists of the *outer boundary* of c'. Summing up we have the simple but often to be used

Theorem. *The probability measure of a subconfiguration c' of c conditioned by the rest of c depends only on the generators in c' and in the outer boundary of c'.*

We then speak of a *Markov random field*, a natural generalization of the concept of a Markov chain. We shall see in Part III that this has important computational consequences for pattern synthesis and inference.

Remark. We shall sometimes define the probabilities on $\mathcal{C}(\mathcal{R})$ for some regularity $\mathcal{R}$. The above construction can, however, also be applied to a configuration space containing also irregular configurations, in which case we speak of *relaxed regularity*: the limit at T or $\epsilon \downarrow 0$ is the *rigid regularity*.

4.6. Deformations. The images in $\mathcal{I}$, the pure images, are mathematical abstractions that may not be exactly (but are in principle) observable. To describe the effect of the sensors we introduce mappings $d \in \mathcal{D}$, the set $\mathcal{D}$ called the *deformation mechanism*, from $\mathcal{I}$ to some space $\mathcal{I}^\mathcal{D}$ of *deformed images*: $d : \mathcal{I} \to \mathcal{I}^\mathcal{D}$. $\mathcal{D}$ expresses the physics or physiology of the sensors among which we mention digital cameras, X-ray cameras, laser radar, and the human visual system; others will be discussed later.

A special case is when $\mathcal{I}^\mathcal{D} = \mathcal{I}$ so that the deformed images preserve the structure of the (pure) image algebra $\mathcal{I}$. We then say that $\mathcal{D}$ is an *automorphic* deformation mechanism. In most cases $\mathcal{D}$ destroys the structure of $\mathcal{I}$, the deformations are *heteromorphic*. Usually $\mathcal{I}^\mathcal{D}$ is then only loosely structured and the d's may lose essential information.

Deformation mechanisms can operate on different levels: they can be given as defined on G, acting on each generator separately, or on $\mathcal{C}(\mathcal{R})$ deforming configurations, or, most commonly, directly on $\mathcal{I}$.

We shall encounter four *types* of deformations: *contrast, background, direct,* and *incomplete* deformations. First, *contrast patterns* where the pure/images I can be identified with a function $I : X \to Y$ from a *background space* X to a *contrast space* Y. For example $X =$

unit square $[0,1]^2 \subset I\!R^2$ and $Y = \{0, 1, 2, \ldots 255\}$ representing 256 gray levels. A contrast deformation d that means a function $I \to f[I(x), x]$.

On the other hand a *background deformation* in this situation could be given by a homeomorphic mapping (note 4.6) $h : X \longleftrightarrow X$ and the contrast image $I = I(x)$ is changed into $I^\mathcal{D} = I[h(x)]$.

Indirect deformations are heteromorphic, so that $\mathcal{I}^\mathcal{D}$ can be of completely different form from $\mathcal{I}$. *Tomographic* deformations (see also section 3.6) belong to this family; an element $I^\mathcal{D}$ is here a function from the set L of line l in X to $I\!R$. For example, if $I(\cdot)$ is a contrast image,

$$I^\mathcal{D}(l) = \int_\ell I(x) dx$$

where the integral is extended overall $l \cap X$ (note 4.7).

The *incomplete* deformations delete information by restricting the observations. For example a contrast image $I = I(x)$ is deformed through

$$I^\mathcal{D}(x); \ x \in M \subset X$$

where M, the *mask* is a proper subset of X. M could be a plane in $X = I\!R^3$.

Deformations can be deterministic or they can be of random nature. The latter will be the case in our example.

Example. Let $\mathcal{D}$ be a noisy channel from the set $\{0, 1\}$ to itself. The image $I(x, y) = 0$ or 1 is deformed into

$$I^\mathcal{D}(x, y) = \begin{cases} I(x, y) \text{ with probability } 1 - \epsilon \\ 1 - I(x, y) \text{ with probability } \epsilon. \end{cases}$$

In section 9.2, Figure 2a, we show an example of this.

4.7. Tasks of pattern theory. Once we have constructed the representations in the form of regular structures we can use them for many purposes. Actually, constructing the representations will turn out to be the hardest part in our endeavor; once they have been built their use will be *derived* by applying general mathematical and computational principles.

The tasks will be grouped into two categories: *synthesis* (or simulation) and *analysis* (or inference). Let us start with synthesis, which is actually the first to be encountered logically in every new application of pattern theory.

4.7.1. Synthesis. Given a regular structure, $\mathcal{C}, \mathcal{I}$, or $\mathcal{P}$, how well does it mimic the real world patterns that we intend to represent? To find an answer to this crucial question we shall *synthesize*, or simulate, the regular structures by an algorithm/program and display them on the computer screen, on a picture or plotter, or other device. If the results look realistic, both in terms of structure and variability, we can proceed to the main tasks. Unfortunately synthesis will often show pattern behavior different from what we intended, and if so, we have to modify our current representation. Experience has shown that it is difficult to predict the pattern behavior from the mathematical model since it may be counterintuitive. Even simple texture patterns can be harder to simulate than one would expect. Synthesis is therefore indispensable as a check on the realism of the mathematical model, but it is of course only a first step in actually using it. That is done by *pattern inference* or *analysis* that comes in the following forms.

4.7.2. Image restoration. The most common, but perhaps not the most interesting, of pattern analysis tasks is that of trying to restore the (pure) image I given an observed one $I^\mathcal{D}$. In other words we want to invert the deformation

$$\mathcal{D}: \mathcal{I} \to \mathcal{I}^\mathcal{D}, \ I^\mathcal{D} = dI; \ d \in \mathcal{D}$$

by a restoration algorithm r

$$r: \mathcal{I}^\mathcal{D} \to \mathcal{I}^*, \ I^* = rI^\mathcal{D}$$

in such a way that I^* is close in some specified sense to I. A reader familiar with statistics will recognize this task as statistical estimation, but with unconventional sample and parameter spaces. Typically these spaces will be high or infinite dimensional so that classical estimation methods can be used only after appropriate modifications.

4.7.3. Pattern recognition. Another task in pattern research that has received much attention is recognition, which means that one wishes to find an algorithm r that attempts to find the pattern class $\mathcal{P}_\gamma$ that I belongs to when $I^\mathcal{D}$ has been observed, or

$$I^\mathcal{D} = dI; \ \mathcal{P}_\gamma^* = rI^\mathcal{D}.$$

4.7.4. Image extrapolation. A less frequent task is when $\mathcal{D}$ is of "incomplete" type as in section 4.6 and we want to extrapolate I from the masked observation $I^\mathcal{D}$. This is related to traditional prediction theory in stochastic processes, most of which is formulated for a one-dimensional index set (time). Here we deal with 2-D or 3-D, however, and the assumptions are also quite different. Some stereological problems belong to D.

4.7.5. Image segmentation. A task of practical interest, especially for texture patterns, is segmentation. This means that given an image, usually of *contrast* type, so that it can be represented by a function $y = y(x), y \in Y =$ contrast space, $x \in X =$ background space. We want to partition it into subsets, where each subset is statistically homogeneous in the sense that it can be described by a single regular structure of relaxed regularity. Therefore one could say that this is the study of heterogeneous patterns but it can also be considered as a special case of pattern recognition.

4.7.6. Image understanding. The most exciting, but also the least developed, of the tasks we will be confronted with is image understanding. This term is meant to incorporate

intrinsic understanding: assuming the pattern representation to be correct, identify its constituent parts in an observed image, and

extrinsic understanding: find where and how the pattern representation is incorrect in an observed image.

For the second part we need an *abnormality* detector that is sensitive to pathological pattern behavior.

We claim that we understand the pattern only if we can analyze it in terms of a given knowledge representation or, alternatively, if we can pinpoint in what respect it differs from what the representation prescribes.

The above list is certainly not complete. Also, it may give the erroneous impression that we are only concerned with cases of a statistical nature. There are many pattern theoretic setups where it is not natural to employ probabilistic/statistical language. If so, the above descriptions have to be modified appropriately.

Chapter 5

Analysis of Open Patterns

The basic concepts for building knowledge representations of pattern ensembles were presented in a general and abstract way in the previous chapter. To fix ideas and clarify the role of these concepts we shall now consider a number of concrete examples. The question of how to organize the computing of the associated regular structures will be postponed to Chapter 8.

5.1. Character strings. To start with the simplest case let us consider configurations consisting of certain strings of characters. Let $G = \{a, b, c, \ldots\}$ be coded by the integers $\{0, 1, 2, \ldots p-1\}$ and let $\omega(g) = 2$ with Σ as directed regularity $\Sigma = $ LINEAR $=$ the set of all finite linear graphs. For $\sigma \in \Sigma$ we enumerate its sites $1, 2, \ldots n$.

We shall at first let bonds carry *full information* in the sense that $\beta_{in}(g) = \beta_{out}(g) = g$ so that bonds determine the generator completely (later on this will not always be the case). A configuration will then look like the one in Figure 1, consisting of a chain of generators hooked together from left to right.

FIGURE 1. LINEAR CONFIGURATION

As an example let us define local regularity by the bond relation

$$\rho(\beta', \beta'') = \begin{cases} \text{TRUE} & \text{if } \beta'' = \beta' \text{ or } \beta'' = \beta' + 1 \\ \text{FALSE} & \text{else} \end{cases}$$

where addition in $\beta' + 1$ is understood modulo p and where the β-values are the integers coding G. For example, if $G = \{a, b, c, d\}$ coded into $\{0, 1, 2, 3\}, p = 4$, we get regular configurations like

$$\begin{cases} bbcddaaa \\ abccd \\ ccccd \\ \ldots \end{cases}$$

Such configurations could be used to describe a system that can be in p different states (or regimes) and where a state can be repeated or jump to the next one, but no larger jumps are allowed.

Now let us add some structure by *positioning* the generators in a *background space* X, for example $X = \mathbb{Z} = \{\ldots - 1, 0, 1, 2, \ldots\}$ consisting of all integers, so that generators consist of a character together with an x-value saying where in X it is positioned. Let bond values still carry full information so that $\beta_{in}(g) = \beta_{out}(g) = g = \{u, v\}$ where u stands for a character and v is an x-value. We shall introduce a new bond relation by a logical conjunction

$$\rho_2((u,v),(u',v')) = \rho(u,v) \bigwedge \mathbf{1}_{v'=v+1}$$

meaning that ρ must be TRUE and $v' = v + 1$. Here we have used the notation $\mathbf{1}_E =$ the *indicator function* of the event E which is TRUE if E occurs.

This leads to a configuration space $\mathcal{C}(\mathcal{R}_{positioned})$ with elements like

$$\begin{cases} b_1 b_2 c_3 d_4 d_5 a_6 a_7 a_8 \\ a_{11} b_{12} c_{13} c_{14} d_{15} \\ c_{-5} c_{-4} c_{-3} c_{-2} d_{-1} \\ \ldots \end{cases}$$

with obvious notation. It is then natural to introduce a similarity group $S = \mathbb{Z}$ consisting of translations

$$\begin{cases} sg = s(u,v) = (u, v+s) \\ g = (u,v), \; s \in \mathbb{Z} \end{cases}$$

so that the first of the above configurations is mapped, with $s = 3$, into

$$b_4 b_5 c_6 d_7 d_8 a_9 a_{10} a_{11}.$$

The similarities just *shift* the string rigidly left or right a certain number of steps on the background space X. The physical interpretation of this could be that x measures time in discrete units and we are dealing with a time homogeneous system.

What is the real distinction between string configurations and positioned string configurations? The regular structure associated with the later ones has $S =$ translation group.

The pattern families $\mathcal{P} = \mathcal{C}(\mathcal{R}_{positioned})/S$ given as a quotient space (note 4.8) then simply consists of the corresponding character strings without positioned information. The pattern aac, for example, contains configurations from the original $\mathcal{C}(\mathcal{R})$ like

$$\begin{cases} a_7 a_8 c_9 \\ a_{-2} a_{-1} c_0 \\ \ldots \end{cases}$$

A pattern has less information than one of its configurations (or images to be more precise, here that distinction is not essential) and the loss of information is described by the similarity group.

We can add still further structure by associating values f_{in}, f_{out}, both from $\mathbb{R}^2$, to the knowledge represented by the generator, and with the new bond relation

$$\begin{cases} \rho_3((u,v,f),(u',v',f')) = \rho_2((u,v),(u',v')) \wedge \mathbf{1}_{f=f'} \\ f = f_{out}, f' = f_{in}. \end{cases}$$

In this way we build more and more information into the regular structure and will get more powerful and detailed representations of the knowledge.

Before leaving character string configurations let us also apply the ideas of section 4.5 to them. Say that the acceptor function, or here matrix, A is given

$$A = (A(x,y); x,y = 0,1,2,\ldots p-1)$$

in such a way that

$$\sum_{y=0}^{p-1} A(x,y) = 1 \text{ for all } x.$$

Then the conditional probability of a configuration $c = \text{LINEAR}(x_1, x_2, \ldots x_n)$ for fixed x_1 is

$$p(c) = A(x_1, x_2) A(x_2, x_3) \ldots A(x_{n-1}, x_n)$$

and the sequence $x_1, x_2, \ldots$ is simply a Markov chain with a transition probability matrix A. For simplicity assume that all strings start with $x_1 = 1$.

This is quite straightforward, but the following is potentially of great importance although we shall describe it only in a very limited context. Let us think of the above probabilities as describing *normal* strings but also occasionally allow for the appearance of abnormality in a string. To make this precise we shall assume, quite arbitrarily and just to illustrate the idea, that a string $a \in \mathcal{C}$ is corrupted by a probability p_{abnorm}. Conditional upon this happening we assume that, substring $x_i, x_{i+1} \ldots x_{i+\ell-1}$ is replaced by a completely random string $y_i, y_{i+1}, \ldots y_{i+\ell-1}$ from $\{0, 1, \ldots p-1\}$ so that we observe

$I^{\mathcal{D}} = (x_1, x_2, \ldots x_{i-1}, y_i, y_{i+1}, \ldots y_{i+\ell-1}, x_{i+\ell} \ldots x_n)$, and that i is randomly chosen from $\{2, \ldots n - \ell\}$. In other words the normal string is perturbed in a chaotic way in a subset of length ℓ.

Our task is to determine 1) if it is likely that a presented string is abnormal and 2) if this is so, where the abnormality is located (see section 4.7.6). To organize automatic abnormality detection in this case let us calculate

$$p_0 = P(\text{normal and } I^{\mathcal{D}} \text{ observed}) = (1 - p_{abnorm})A(x_1, x_2)A(x_2, x_3)\ldots A(x_{n-1}, x_n).$$

Also for $i = 2, \ldots n - \ell$

$$p_i = P(\text{abnormality starting at site } i \text{ and } I^{\mathcal{D}} \text{ observed}) =$$
$$= p_{abnorm} \frac{p^{-\ell}}{n - \ell - 1} \sum A(x_1, x_2) \ldots A(x_{i-2}, x_i) \ldots A(x_{n-1}, x_n)$$

where the summation is over $x_i, x_{i+1} \ldots x_{i+\ell-1}$. But that summation corresponds to forming the power $A^{\ell+1}$ of the acceptor matrix so that we can write

$$p_i = p_{abnorm} \frac{p^{-\ell}}{n - \ell - 1} A(x_1, x_2) \ldots A(x_{i-2}, x_{i-1})$$
$$A^{\ell+1}(x_{i-1}, x_{i+\ell})A(x_{i+\ell+1}, x_{i+\ell+2})\ldots A(x_{n-1}, x_n).$$

We then find

$$q_0 = P(\text{abnormality}|I^{\mathcal{D}} \text{ observed}) = \frac{P(\text{abnormality and } I^{\mathcal{D}} \text{ seen})}{P(I^{\mathcal{D}} \text{ observed})}$$
$$= \frac{\sum_{i=2}^{n-\ell} p_i}{p_0 + \sum_{i=2}^{n-\ell} p_i}.$$

If this probability exceeds some threshold value $p_{critical}$ we decide that the string is abnormal. Having made that decision we can also estimate the beginning site of the abnormal subset by finding the i-value that maximizes p_i. A more conservative decision rule would be to determine a set of i-values for which p_i is big. We shall study in section 8.2 how this can be implemented computationally to give us code for the abnormality detector.

5.2. Operator strings. Consider now instead p linear differential operators, say of second order,

$$L_k f = \frac{d^2 f}{dx^2} + a_k(x)\frac{df}{dx} + b_k(x)f; \quad k = 0, 1, \ldots p - 1;$$

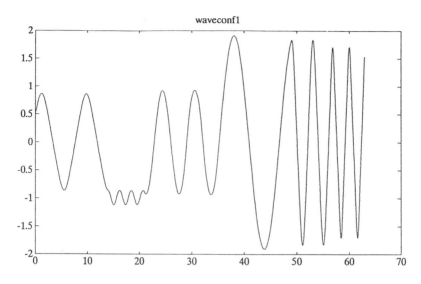

FIGURE 2. WAVE FORM

operating on functions $f(x)$, x real. Consider solutions of $L_k f = 0$ on an interval $(x, x+\ell)$ where x and ℓ are integers. If we specify the values $x, f(x), f'(x)$ the solution is unique; in particular it determines also the values $f(x+\ell), f'(x+\ell)$. Therefore we can write

$$(f(x+\ell), f'(x+\ell)) = \varphi_k(x, f(x), f'(x))$$

with some function φ_k associated with the operator L_k and the chosen (fixed) length ℓ.

The new regularity $\mathcal{R}' = <G', S, \Sigma, \rho'>$ will be chosen as G' consisting of solutions to $L_k f = 0$ over some interval $(x, x+\ell)$ and arity two with the bonds

$$\begin{cases} \beta_{in}(g) &= (k, x, f_{in}) \\ \beta_{out}(g) &= (k, x, f_{out}) \end{cases}$$

and ρ' shall mean that $k' = k$ or $k+1 \pmod{p}$, $x' = x + \ell$, and $f_{out} = f_{in}$ where f_{out}, f_{in} stand for the value of f and f'. We use essentially the same S by letting an element s translate the x-values by $s\ell$ (not just ℓ). Compare this with the last regularity of the previous section.

With this setup we see that we have established a configuration homomorphism (see section 4.2) that preserves structure but changes strings of characters into strings of operators or the functions that solve the corresponding equations. With a choice of operators that will be introduced in section 8.1 the string 0 0 1 2 3 4 4 0 1, $p = 5$, is homomorphic to the function graphed in Figure 2.

One can see how different regimes follow each other: a succession of wave forms defined by differential operators are continuously (including the first derivative) concatenated to each other.

Remark. A relaxed form of this rigid regularity would be obtained if we let transitions from one character to another be given by a Markov chain that allows bigger jumps than one with positive probability. One could also replace the differential equations by

$$L_k f = e = \text{white noise}$$

which would produce deformed wave forms.

5.3. Finite state languages. Let us now interpret the finite state grammars, FS, from section 2.2 in terms of regular structures. We shall choose as generators the rewriting rules (not the words!) expressed diagrammatically as

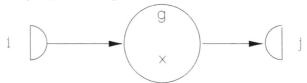

for the rewriting rule $i \xrightarrow{x} j$. Here i, j are the syntactic variables from V_N and x is the word from V_T that is written when the rule is applied. In other words $\omega(g) = 2$, $\beta_{in}(g) = i$, $\beta_{out}(g) = j$ and we select the bond relation $\rho = \text{EQUAL}$: $\rho(\beta, \beta') = \text{TRUE}$ iff $\beta = \beta'$. The bond value space is here $B = V_N$.

The connection type is that of (directed) linear strings, $\Sigma = \text{LINEAR}$, so that no cycles occur and we are dealing with open patterns. Any regular configuration with $j_1 = i_2, j_2 = i_3, \ldots$ can be written as the configuration diagram

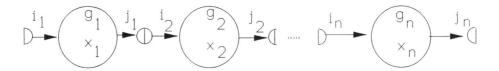

and represents a *grammatical phrase* $x_1, x_2, \ldots x_n$. Note that it is not always true that such a phrase has a unique parse, the identification rule R is not necessarily bijective. In the special case when it is we speak of an *unambiguous* grammar.

We shall be particularly interested in grammatical *sentences*: the word strings that correspond to regular configurations starting with $i_1 = 1$ and ending at $i_n = F$, a given final state. The set of these sentences constitutes the *language* generated by the grammar. This is made a metric pattern by the second structure formula of the type given in section 4.5 (2) with, for the time being,

$$A(\beta, \beta') = \rho(\beta, \beta') \text{ interpreted as 1 or 0}$$

and some weight function $Q(\cdot)$. This means that we get a Markov chain with transition probabilities

$$P(g \to g'|g \text{ fixed}) = \frac{Q(g')}{\Sigma Q(g'')}$$

if $\beta_{out}(g) = \beta_{in}(g')$ and the sum is over all g'' with $\beta_{in}(g'') = \beta_{out}(g)$. In all other cases the transition probability is zero.

This means that the support of the probability distribution is contained in $\mathcal{C}(\mathcal{R})$: positive probability occurs if the configuration is regular. It may be that we are instead interested in allowing irregular configurations to have positive probability relaxing the regularity as mentioned in section 4.5. We would then allow the acceptor function $A(\cdot, \cdot)$ to take other values than 0 or 1. We shall see how to implement this on the computer in section 8.3 where an example is also given of a simple but illustrative FS grammar.

5.4. Context free languages. We shall now consider CF grammars that formalize the discussion in section 2.2. The vocabulary $V = V_N \cup V_T$ now has the non-terminal vocabulary V_N consisting of rewriting rules of the form

$$g : x \to x_1 x_2 \cdots x_m; \; x \in V_N, x_k \in V.$$

The rewriting rules, the generators, can therefore be visualized as in Figure 3 with $\omega_{in}(g) = 1$, $\omega_{out}(g) \geq 1$, $\beta_{in}(g) \in V_N$, $\beta_{outj}(g) \in V$.

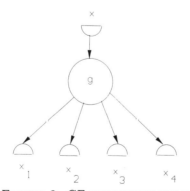

FIGURE 3. CF REWRITING RULES

With ρ = EQUAL, Σ = TREE we get regular configurations that could look like the one in Figure 4 where we have indicated the upper inbond β and the lower outbonds $\beta_1, \beta_2, \beta_3, \beta_4$: the external bonds with terminology from section 4.2. This defines $\mathcal{C}(\mathcal{R})$ except that we have not defined the similarities. One way of doing this that is sometimes natural is to introduce a permutation group S_P on V_T, such that word classes are invariant with respect to S_P, and to define $sg, s \in S_P$ by permuting the terminals (if any) appearing as outbonds of g according to s.

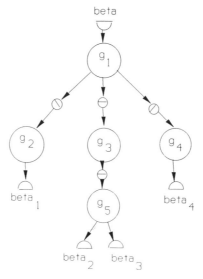

FIGURE 4. REGULAR CF CONFIGURATION

We shall identify two regular configurations $c_1, c_2 \in \mathcal{C}(\mathcal{R})$ if $\text{ext}(c_1) = \text{ext}(c_2)$. This identification rule R defines the images, here called *grammatical phrases*, and the corresponding image algebra $\mathcal{T}$. In particular, complete derivation trees whose inbond is the initial state S and whose outbonds are all terminals are called *grammatical sentences*. In general we cannot claim that each phrase (or sentence) is generated by a single configuration; ambiguity can occur when we parse.

Probabilities on $\mathcal{C}(\mathcal{R})$ are introduced by setting $A(\cdot, \cdot) = \rho(\cdot, \cdot)$ and with a Q-vector, as in equation (2) of section 4.5. One could also use more relaxed regularity, allowing for common grammatical mistakes, say in colloquial language, by letter $A(\cdot, \cdot)$ taking other non-negative values than just 0 and 1. In section 8.3 we shall experiment with this language model.

5.5. Curve images. Now let us turn to geometric patterns and apply the principles of Chapter 4, in particular that of deformable templates. Start with a single curve template $c^0 = \sigma(g_1^0, g_2^0, \ldots g_n^0)$ where the g_i^0 are line segments $\overrightarrow{(z_i, z_{i+1})}$ in the plane so that $G = \mathbb{R}^4$. Let $\omega(g_i^0) = 2$ with

$$\begin{cases} \beta_{in}(g_i^0) = z_i \\ \beta_{out}(g_i^0) = z_{i+1} \end{cases}$$

so that the bond value space $B = \mathbb{R}^2$. We shall choose the connection type LINEAR and bond relation $\rho(\beta_{out}, \beta_{in}) = $ TRUE if $\beta_{out} = \beta_{in}$, $\rho = $ EQUAL.

This means that a configuration is a sequence of sides and we shall identify two configurations if they represent the same polygon as a set of points. It may appear that in this case R is just the identity relation so that the concepts of image and configuration coincide. This is not so, however, since a side of a polygon can be represented by a concatenation of parallel sides. This innocuous observation will have unpleasant consequences.

Here we shall operate with two similarity groups, S_1 and S_2, where S_1 is the translation group in the plane. Of course the vertices of the polygon can be written as cumulative sums

$$V_i = V_0 + v_0 + v_1 + \cdots + v_{i-1}$$

where V_0 is one vertex and the v_i's are vectors $z_{i+1} - z_i$. Line segments modulo S_1 means vectors, and a curve pattern (in the sense of section 4.4) consists of all possible translates of a given curve. Ideas like this are as old as Euclid, but with another group, where the notion is called congruence.

Now we shall choose one or several curves, represented by $\sigma(g_1^0, g_2^0, \ldots g_n^0)$, and deform them probabilistically. To make this precise we introduce another similarity group $S_2 = 0(2) = $ the set of rotations of vectors in the plane and where we can also let s stand for the rotation it represents. Hence $s \in [0, 2\pi)$ and we shall pick g_i^0 as unit vectors of orientation angle φ_i. Thus the deformed template can be written as

$$\begin{cases} c = \sigma(s_1 g_1^0, s_2 g_2^0, \cdots s_n g_n^0) \\ s_i g_i^0 = \text{unit vector with orientation } \varphi_i + e_i \end{cases}$$

where the random group elements (here just angles) e_i will be given a joint probability density as in section 4.5 where

$$A(e', e'') = \frac{1}{Z(e')} exp[a\ cos(e' - e'') + b\ cos\ e'']$$

where $Z(e')$ has been chosen to make $A(e', e'')$ a conditional probability density

$$\int_0^{2\pi} A(e', e'') de'' = 1$$

so that $Z(\cdot)$ can be expressed as an elliptic integral.

The joint density $p(e_1, e_2, \ldots)$, which will be called the *multivariate von Mises density*, has a coupling strength parameter a and a parameter b that measures the peakedness of the distribution. The question is then the usual one in pattern synthesis: how do we simulate the distribution over $\mathcal{C}(\mathcal{R})$ which here is given in terms of the $s_i = \varphi_i + e_i$.

But this is straightforward. Indeed, because of the multiplicative form of $p(e_1, e_2, \ldots)$ the e_i's form a Markov process taking values on $[0, 2\pi)$. We just fix a value e_0, somewhat arbitrarily put $= 0$, simulate $A(0, \cdot)$ to get e_1, then simulate $A(e_1, \cdot)$ to get e_2, and so on.

With a template consisting of several rays as in Figure 5(a) we get the mild deformation in (b) and the severe one in (c); see section 8.4 for the computation. Note that the strategy for synthesis was close to the one in sections 5.3 and 5.4: we exploit the Markov chain structure and take one step at a time. This is possible because we are still dealing with open patterns; the connectors have no closed cycles. We shall return to deformable templates in the same geometric setting, but with closed cycles, in Chapter 6.

Chapter 5: Analysis of Open Patterns

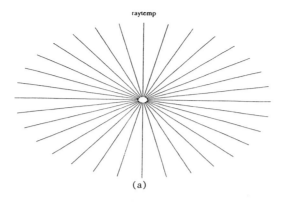

(a)

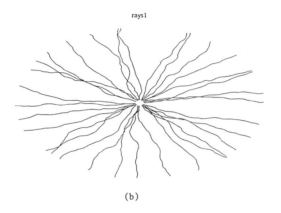

(b)

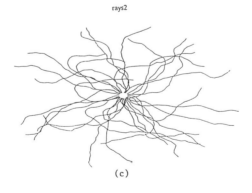

(c)

FIGURE 5

Chapter 6

Analysis of Closed Patterns

The mathematical understanding of closed patterns is more challenging than for the open ones; this is due to the more intricate dependence architecture. But it is also more fun to confront these difficulties.

We are now forced to use iterative schemes with their several drawbacks (much computing, difficult to know when to stop, etc.), but no direct methods are presently available that can be implemented on conventional digital architecture, whether sequential or parallel. There is a vague belief among a few practitioners that this will change when we have a powerful analog technology on VLSI or charged coupled devices, but this is in the future (note 6.1).

6.1. The Ising model. One of the simplest and most celebrated probabilistic models with cycles in the connector graph is the *Ising model*. Originally introduced in physics for the study of magnetism, it has been used for representing 2-D picture ensembles with some success, at least for textures.

Consider a rectangular lattice of size nr rows and nc columns. At each site x, y of the lattice is placed a generator $g \in G = \{0, 1\}$ with full bond information, $\beta_j(g) \equiv g$ for $j = 1, 2, 3, 4$ so that we have arity $\omega(g) = 4$. The acceptor function is then just a 2×2 matrix that we will denote by

$$A = \begin{pmatrix} 1 & b \\ b & c. \end{pmatrix}$$

There is no loss of generality in choosing the upper left entry as 1 since any other choice would be automatically compensated for when dividing by the partition function Z in the second structure formula from section 4.5.

The meaning of b is that small b-values tend to produce cohesive pictures: large blocks of 0's alternate with large blocks of 1's. A large b-value will tend to produce checker-like pictures. Large c-values will lead to lots of 1's in the picture and vice versa for small c's. The case $b = c = 1$ gives a completely chaotic picture ensemble; all configurations are equally likely.

The probability density in section 4.5 is, it may appear, of unconventional type but there is nothing peculiar about it. For any configuration c we can compute the probability $p(c)$ once we have the normalizing constant Z, the partition function. But here is the rub: how are we to find Z? The best we could hope for is a formula for Z in terms of b and c. It is true that for special cases this can be done (note 6.2) but in general it cannot. And even if we could find Z how could we possibly use the calculated values $p(c)$, for $|\mathcal{C}| = 2^{1000}$ or worse? To solve this conundrum we shall have to perform some mathematical acrobatics with a long history (note 6.3). The reader is warned that it is trickier than the earlier methods.

Say that we consider a site i in the connector graph $c = \sigma(g_1, \ldots g_i \ldots g_n)$ and look at the environment $env(g_i)$ (see section 4.1);

$$\begin{cases} env(g_i) & = (\beta'_1, \beta'_2, \ldots \beta'_\omega) \\ B_v(g_i) & = (\beta_1, \beta_2, \ldots \beta_\omega) \end{cases}$$

where $B_v(g)$ stands for the set of bond values from g_i as in Figure 1. Here $\omega(g_i) = 4$.

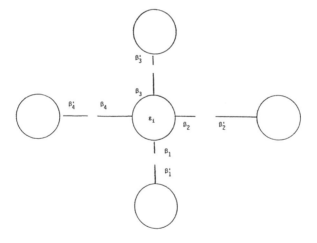

FIGURE 1

Now we are going to let g_i take all values $g \in G$ and compute the conditional probability $p(g|env)$ of getting g when the environment is fixed. Using the second structure formula in section 4.5 we get

$$p(g|env) = \frac{P(g \text{ and } env)}{P(env)} \propto \sum p(c)$$

summed over all c with given $g_i = g$ and $env(g_i)$. The denominator above is just a constant, a proportionality factor, so that the structure formula reduces to

$$p(g|env) \propto \sum \prod_{<k,k'>} A[\beta_{j_1}(g_{i_1}), \beta_{j_2}(g_{i_2})]$$

summed over the same configuration c as just mentioned. This complicated expression simplifies dramatically since we "sum out" everything except for the generator g and the bonds β'_j of env.

Therefore, for the situation in Figure 1,

$$p(g|env) \propto A[\beta_1(g), \beta'_1]A[\beta_2(g), \beta'_2] \times$$
$$\times A[\beta_3(g), \beta'_3]A[\beta_4(g), \beta'_4] = q(g|env).$$

The proportionality constant left out is just

$$\frac{1}{\sum_{g \in G} q(g|env)}.$$

Several things should be observed. First, the troublesome partition function Z no longer appears in the expression for $p(g|env)$. Second, the number of multiplications, additions, divisions is only $3 \times |G| + |G| - 1 + |G| = 5|G| - 1$ so that for moderate size $|G|$ of the generator space the computing is almost instantaneous.

Note. If the connector graph σ has sites of much larger arity the computation will not be so fast: locality of the graph helps!

Now let us describe one of many iterative simulation methods for pattern analysis when the patterns are closed.

Step 1. Initialize the configuration $c = c(1)$, at time $t = 1$ in some way. In principle this is arbitrary but to speed up convergence one may be able to find an especially good initialization.

Step 2. Select a site i, perhaps at random, or by scanning the lattice left-right, up-down, as a TV picture. To fix ideas it will be assumed that it is chosen at random.

Step 3. Compute $\beta'_1, \beta'_2, \ldots$ in the environment of g_i and the conditional probabilities $p(g|env)$.

Step 4. Randomize the choice of g, *update g_i*, using this probability distribution which can be done quickly unless $|G|$ is large.

Step 5. Replace the old g_i at site i by the new g.

Step 6. Go to Step 1 until a sufficient number of iterations have been made, then stop.

This simulation scheme is known as *stochastic relaxation* (and has other names) and gives a solution to our problem:

Theorem. *If all $A > 0$, strictly positive, the sequence $c(0), c(1), c(2) \ldots c(t)$ of random configurations will converge in probability to the desired probability $p(\cdot)$ that we started with.*

Remark 1. Note that we did not claim that $c(t)$ tends to some fixed c as $t \to \infty$; that will not happen. Convergence in probability should be understood to mean

$$\lim_{t \to \infty} P[c(t) = c] = p(c), \text{ all } c.$$

Remark 2. We did not say what a "sufficient" number of iterations really means, because we do not know a lot about it. Much empirical experience about this is available, some of it is controversial, but firm analytical understanding of what speed of convergence to expect has only become available during the last few years.

To prove the theorem let us observe that the sequence of configurations $c(0), c(1), (2), \ldots$ forms a path through $\mathcal{C}$ and that when we randomize $c(t+1)$ we use the knowledge of $c(t)$ but not of the earlier ones $c(t-1), c(t-2), \ldots$. In other words $c(t)$ forms a Markov chain in the state space $\mathcal{C}$ (note 6.4).

Also note that since all $A(\beta, \beta')$ are strictly positive it is possible to go from any c to any c' in a finite number of steps with positive probability. But then an important theorem in Markov chains (note 6.4) says that the probability distribution p_t of $c(t)$ converge to a limit p_∞. Furthermore p_∞ is an *equilibrium distribution* in the sense that if we select a c at random from $\mathcal{C}$ according to the probabilities $p_\infty(\cdot), p(c) = p_\infty(c)$, and apply our random updating then the new configuration c' will also satisfy $p(c') = p_\infty(c')$. The equilibrium probability p_∞ is also unique.

But then we know that $p[c(t) = c] \to p_\infty(c)$ and also that $p_\infty(\cdot)$ coincides with the density given by the second structure formula. Indeed, if c is randomized according to the latter, the new c' will have the same probability (of course c' may differ from c but their probabilities are the same); we have an equilibrium distribution. That means the limit p_∞ is exactly the one we wanted to simulate so that the statement in the theorem is true.

Q.E.D.

For the Ising model we show one synthesis in Figure 2 with $b = .05, c = 1$ in the A-matrix. The small value of b produces a cohesive picture with large regions of 0's and others of 1. The code for this will be given in section 9.1.

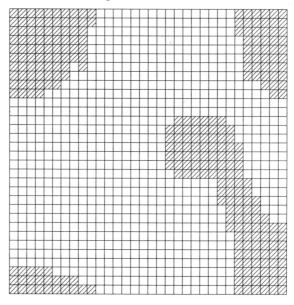

FIGURE 2. ISING IMAGE

Now we turn to the inverse problem, *pattern analysis*, and do it in the form of restoring (see section 4.7.2) a deformed image $I^{\mathcal{D}} = \mathcal{D}I$ (image and configuration mean the same thing in this context), where I is an Ising image. Say that $\mathcal{D}$ is just a contrast deformation with

$$P(c^{\mathcal{D}}(x,y) = v|c) = \begin{cases} 1 - \epsilon \text{ if } v = c(x,y) \\ \epsilon \text{ else.} \end{cases}$$

In signal processing this means that $\mathcal{D}$ is a symmetric binary noisy channel with error probability ϵ.

One natural approach here (there are several others) is to simulate the posterior density, which we can write, using Bayes' theorem (note 6.5), as

$$p(c|c^{\mathcal{D}}) = \frac{p(c, c^{\mathcal{D}})}{p(c^{\mathcal{D}})} = \frac{p(c)L(c^{\mathcal{D}}|c)}{p(c^{\mathcal{D}})} \propto p(c)L(c^{\mathcal{D}}|c).$$

Here $L(c^{\mathcal{D}}|c)$, the *likelihood function*, is the conditional probability of observing $c^{\mathcal{D}}$ when c is the true configuration. The simple nature of $\mathcal{D}$ makes it easy to get L

$$L(c^{\mathcal{D}}|c) = \prod_{x,y} \epsilon^{e(x,y)}(1-\epsilon)^{1-e(x,y)}$$

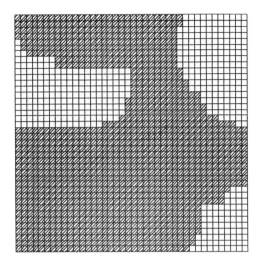

FIGURE 3. ISING IMAGE

where $e(x, y) = 1$ if an error occurs at site (x, y) so that $v = c(x, y)$; otherwise $e(x, y) = 0$.

This is going to cause few changes in the synthesis method to transform it into an analysis program. Indeed, the posterior probability density is also of multiplicative form, now

$$p(c|c^{\mathcal{D}}) = \frac{1}{Z} \prod A[c(x, y), c(x', y')] \prod (1 - \epsilon)^{e(x,y)} \epsilon^{e(x,y)}$$

where the first product is over all pairs of neighboring sites (x, y) and (x', y') on the σ-graph and the second over all sites. But then we can apply the reasoning behind stochastic relaxation with the minute difference that instead of having a $p(g|env)$ given as the product of 4 A-values we also get a contribution from the second product above and get (recall that here $\beta_j(g) \equiv g$)

$$p(g|env, c^{\mathcal{D}}) \propto A(g, \beta_1')A(g, \beta_2')A(g, \beta_3')A(g, \beta_4')(1 - \epsilon)^e \epsilon^{1-e}$$

where $e = 1$ if $g = c^{\mathcal{D}}(x, y)$ and 0 otherwise.

After normalizing these numbers into a probability vector over the generator space G, we update and behave just as for synthesis. But instead we encounter another question: having simulated the posterior $p(c|c^{\mathcal{D}})$ what are we going to do with the result c^*? Sometimes it is good enough just to use the result c^* as a restoration. The Ising image in Figure 3, for example, that corresponds to $b = .01, c = 1$ distorted with $\epsilon = 25\%$ into Figure 4, which leads to the restoration $c^* = I^*$ in Figure 5 and is quite successful. Remember that the algorithm starts from Figure 4 and does not know Figure 3 (note 6.6).

118 *Chapter 6: Analysis of Closed Patterns*

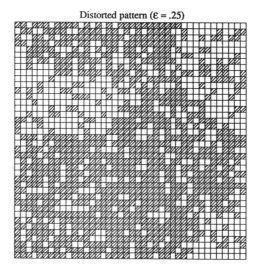

FIGURE 4. DEFORMED ISING IMAGE

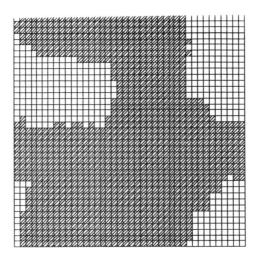

FIGURE 5. RESTORED ISING IMAGE

Remark. Figure 5 was obtained after only three full sweeps, so that for this 40×40 lattice we needed only 4800 single updates. Other experience has led us to believe, in contrast to what was just said, that hundreds or thousands of sweeps are sometimes needed. This question is controversial and the reader may wish to experiment to learn more about it.

Instead, one could run the restoration N times, resulting in $c_1^*, c_2^*, \ldots c_N^*$ and sort of average them by forming

$$c^{**}(x,y) = \begin{cases} 1 \text{ if most of } c_1^*(x,y), c_2^*(x,y), \ldots c_N^*(x,y) \text{ are } = 1 \\ 0 \text{ else.} \end{cases}$$

This type of *majority logic* seems to work well. An advantage of this is that the c^*-sample also tells us something about how sure we can be about our restoration; this can be of crucial importance for making decisions based on the images of more practical use (e.g. medical images) to be discussed later on.

Another possibility that is very appealing is *simulated annealing* (note 6.7), but this requires more mathematical background than we have assumed for this book.

Still another technique is *iterated conditional mode*, ICM (note 6.8), which differs in being deterministic (not random). The idea is not to randomize the random generator g_i with probabilities $p(g_i|env)$ but to pick the g_i-value that makes the probability as big as possible, the mode of the distribution. It seems to be quite fast; the drawback is that one cannot make precise claims, as in the Theorem, about what it converges to.

6.2. Boundary patterns. The patterns generated in the previous section are weakly structured; the Ising model represents only the limited knowledge that the images will be cohesive. White pixels will tend to be neighbors of white ones, and black pixels of black ones. To build a bit more knowledge into the representation we shall now choose a generator space that also controls the (random) behavior of boundaries between white and black regions. To do this it is not enough to have one generator for "inside," another for "outside"; we also need generators for boundary elements.

But a boundary element can be straight, it can turn left, or it can turn right. Let us therefore start with an (initial) generator space G_0 consisting of the ones shown in Figure 6; the meaning of the bond values will become clear when we discuss the bond relation ρ and the acceptor function A.

Here we only allow $90°$ turns (left, right) since the arity $\omega(g) = 4$ only. We must, however, also rotate the above g's by $90°, 180°, 270°$ to get all possible orientations. For $g = 0$ and 1 this is redundant, since only the same g's would result. If we carry out these rotations we get an (extended) generator space G of size $|G| = 2 + 4 \times 3 = 14$ and we represent it in Table 1. The generators are enumerated $g = 0, 1, \ldots 13$ and the bond directions by $j = 1, 2, 3, 4$. Check with Chapter 3 and section 9.2.

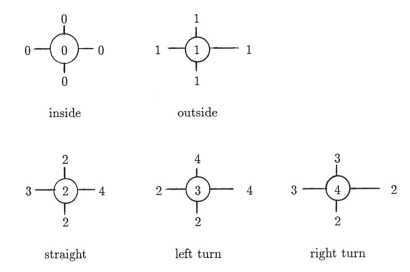

FIGURE 6. GENERATOR SPACE

The couplings between bond values are motivated as follows. Inside elements should tend to stick together, $\rho(0,0) = 1$. Also outside elements, $\rho(1,1) = 1$, but inside and outside should be separated by boundary generators, $\rho(0,1) = 0$. A straight boundary element could continue $\rho(2,2) = 1$ but $\rho(2,0) = \rho(2,1) = 0$. To the left of $g = 2$ we want inside elements $\rho(3,0) = 1$, etc. This is completed in Table 2. Here we have the bond value space $B = \{0, 1, 2, 3, 4\}$ which should be compared to $|G| = 14$.

Chapter 6: Analysis of Closed Patterns

Table 1

$g\backslash j$	1	2	3	4
0	0	0	0	0
1	1	1	1	1
2	2	4	2	4
3	4	2	4	2
4	2	4	2	4
5	4	2	4	2
6	2	4	4	2
7	2	2	4	4
8	4	2	2	4
9	4	4	2	2
10	2	2	3	3
11	3	2	2	3
12	3	3	2	2
13	2	3	3	2

Table 2. Bond relation

$\beta\backslash\beta'$	0	1	2	3	4
0	1	0	0	1	0
1	0	1	0	0	1
2	0	0	1	0	0
3	1	0	0	0	0
4	0	1	0	0	0

Remark. The reader may wonder why we did not replace $\beta = 3$ with 0 and $\beta = 4$ with 1. Experimenting with such a setup using the code in section 9.2 will be instructive: the "boundaries" will not really be boundaries. This is typical: constructing the generator space is not immediate! It can be tricky.

We are, as usual, not interested in quite rigid regularity so we introduce $A(\cdot,\cdot)$ as a 5×5 matrix with entries close to those of ρ: for example,

$$A(\beta,\beta') = \rho(\beta,\beta') + \delta$$

where δ is a small positive constant. We shall use A as in (2) of section 4.5 with a Q-vector that attributes various weights to the different g-values. This can be done almost as in the beginning of section 6.1 except that the Q-value appears in

$$p(g|env) \propto A[\beta_1(g), \beta_1']A[\beta_2(g), \beta_2']A[\beta_3(g), \beta_3']A[\beta_4(g), \beta_4']Q(g).$$

To choose Q note that we have two generators in G for inside/outside, of which we want many occurrences in the image, but twelve g's for boundaries, of which we want much fewer in the image. Therefore the first two entries of Q should be much larger than the following ones. In section 9.2 we shall experiment with this regular structure that knows a bit more than the Ising model about the image ensembles. Note, however, that the knowledge is still only local, the model still does not know anything about global properties like circles, triangles, not to mention really complex ones like stomach shapes.

6.3. Deformable templates. In the two previous sections we described some weak and local but useful knowledge representations of plane patterns. They could only express some vague qualitative properties, the first one cohesiveness of binary patterns, the other one boundary behavior. To build more precise knowledge, not just of local type, but describing the global properties of the shape, we shall exploit the deformable template idea from section 4.2. To fix ideas let the template be a closed curve

$$\begin{cases} c_{temp} = \text{CYCLIC}(g_1^0, g_2^0, \ldots g_n^0) \\ g_i^0 = \text{line segment}(\overrightarrow{z_1, z_2}) \in \mathbb{R}^4 \\ \beta_{in}(g_i^0) = z_1, \beta_{out}(g_i^0) = z_2, B = \mathbb{R}^2 \\ \rho = \text{EQUAL}. \end{cases}$$

With some similarity group $S: G \to G$ we shall deform c_{temp}

$$c_{temp} \to \text{CYCLIC}(s_1 g_1^0, s_2 g_2^0, \ldots s_n g_n^0)$$

so that the new generators are $s_i g_i^0$. We could let S consist of scaling and translations so that

$$sg_i^0 = s \begin{pmatrix} g_{ix}^0 \\ g_{iy}^0 \end{pmatrix} = c \begin{pmatrix} \cos\varphi & \sin\varphi \\ -\sin\varphi & \sin\varphi \end{pmatrix} \begin{pmatrix} g_{ix}^0 \\ g_{iy}^0 \end{pmatrix} = \begin{pmatrix} u & v \\ -v & u \end{pmatrix} \begin{pmatrix} g_{ix}^0 \\ g_{iy}^0 \end{pmatrix}$$

where $c > 0$ is a scaling parameter, φ a rotation angle and

$$\begin{cases} u = c\ cos\ \varphi \\ v = c\ sin\ \varphi. \end{cases}$$

Let us say that we are not interested in location so that all is understood modulo the translation group. But line segments modulo translations are the same as vectors, which is the reason why we represented g_i^0 as a two-vector $\begin{pmatrix} g_{ix}^0 \\ g_{iy}^0 \end{pmatrix}$ above. We are dealing with patterns rather than individual images (compare section 4.4).

How are we going to randomize the u_i's and v_i's to produce smooth but variable deformations? If we made them stochastically independent the configurations would look too chaotic; we must couple them by suitable acceptor functions. Let us do this for the v-sequence and introduce

$$A(v, v') = exp - \frac{1}{2\sigma^2}(v - av')^2$$

where $a = 0$ means no coupling but larger a-values will make the v's depend upon each other. The second structure formula will give the joint probability density for the v's as (with the cyclic convention $v_{n+1} = v_1$, σ =CYCLIC)

$$p(v) = \propto \prod_{i=1}^{n} exp - \frac{1}{2\sigma^2}(v_{i+1} - av_i)^2 = exp - \frac{1}{2\sigma^2}Q$$

with the quadratic form

$$Q = \sum_{k,\ell} m_{k\ell} v_k v_\ell = v^T M v$$

in terms of the column vector v of the v's and the symmetric matrix M as

$$M = \begin{pmatrix} 1+a^2 & -a & 0 & 0 & \cdots & 0 & -a \\ -a & 1+a^2 & -a & 0 & \cdots & 0 & 0 \\ \cdots & & & & & & \\ -a & 0 & 0 & 0 & \cdots & -a & 1+a^2 \end{pmatrix}$$

Note that M has entries that are constant along its diagonals; such a matrix is called Toeplitz. Also that the quadratic form is non-negative definite, since it is a sum of squares, in the sense that

$$v^T M v \geq 0 \text{ for all } v \neq 0.$$

Can it take the value zero? Then there must be a v-vector such that

$$\begin{cases} v_2 - av_1 = 0 \\ v_3 - av_2 = 0 \\ \cdots \\ v_1 - av_n = 0 \end{cases}$$

implying that $v_1 = a^n v_1$ which can be excluded if we make $a \neq 1$, as will be done in the following.

But such a positive definite matrix M has a square root, say N (note 6.9), also symmetric, meaning $N^2 = M$. Let us introduce the inverse $F = N^{-1}$ which is possible since M is non-singular (positive definite) so that N must also be non-singular: $0 \neq det(M) = det(N^2) = [det(N)]^2$, and make the change of variables

$$v = Fw, w = Nv.$$

Then, in terms of w's our joint probability density becomes proportional to

$$exp - \frac{1}{2\sigma^2} v^T M v = exp - \frac{1}{2\sigma^2} v^T N^2 v = exp - \frac{1}{2\sigma^2} w^T w.$$

But $w^T w = \sum w_i^2$ so that the w's are stochastically independent and all $N(0, \sigma^2)$. This is easy to simulate!

Now we do almost the same for the u's (but with new randomness), but remember that the identity element of the similarity group S is the identity matrix $u = 1, v = 0$. Since we do not want wild deformations we therefore make the u's have mean 1, rather than zero, so that we can simulate the u's as the v's, afterwards adding 1 to all its components.

Only one difficulty remains. The template was closed, which means the vector equation

$$\sum_{i=1}^{n} g_i^0 = 0.$$

Both $x-$ and $y-$components of the vector sum are zero. This will not necessarily hold for the new generators $g_i = s_i g_i^0$. A simple way to rectify this (better ones exist) is to adjust to

$$\bar{g}_i = g_i - \frac{1}{n} \sum_{i=1}^{n} g_i.$$

A circle deformed in this way is shown in Figure 7. This looks promising!

Remark 1. If σ is very large the curve produced may intersect itself. In some applications this can be physically meaningful (overlap), in others it cannot. Experience has shown that this is not practically worrisome.

Remark 2. More important is the fact that synthesis is direct, not iterative as in sections 6.1 and 6.2. This is attractive but unfortunately not typical for closed patterns. We shall therefore have to return to this problem but with iterative schemes, in particular for pattern inference.

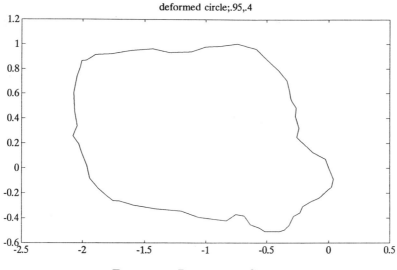

FIGURE 7. DEFORMED CIRCLE

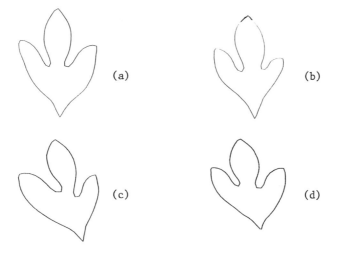

FIGURE 8. SIMULATED LEAF SHAPES

Simulating leaf shape by the same method gives us the four simulated sassafras boundaries in Figure 8. This should be compared with the real boundaries in Figure 28 of section 3.4 (see note 6.10).

The template was a curve, a one-dimensional object situated in 2-D. More powerful knowledge representations are needed if there is internal structure, for example a scan of a brain where not only surfaces are important but also structures inside. This is currently a hot research topic, but because of its advanced mathematical nature we can only describe the simplest case.

Say we are in 2-D with the background space $= [0,1] \times [0,1] =$ the unit square; we use coordinates x, y for the points in the square. But here we have to make a decision that will influence the construction of knowledge representations. Many popular sensors produce digital pictures in the form of rectangular arrays, matrices. The real objects exist in a continuum, in our case $\mathbb{R}^2$, and the fact that the observed picture is discrete is just a technological artifact. It seems natural to build the representation of the image ensemble in the continuum that naturally admits the usual transformation groups: not only translations but also rotations and scaling. Actually, and this will turn out to be crucial, it also admits *homeomorphic mappings*, that is bijective mappings, continuous in both directions. We already made this choice tacitly in the first half of this section, but we meet some new challenges and we shall at least sketch the basic ideas behind this approach. Say that h is a map

$$h : [0,1] \times [0,1] \longleftrightarrow [0,1] \times [0,1]; \; h \in H$$

from some family H of homeomorphic mappings. We use the notation

$$h(x,y) = [x + u(x,y), y + v(x,y)]$$

where $u(\cdot,\cdot), v(\cdot,\cdot)$ are the horizontal and vertical displacement fields.

Given a template $I_{temp}(\cdot,\cdot)$ of contrast type so that it is a function that takes positive values, the gray levels, we shall deform I_{temp} by h via the background deformation

$$I(x,y) = I_{temp}(x + u(x,y), y + v(x,y)).$$

If we define the probability distribution of the u, v-fields this will induce a probability distribution for the image I. Before we do this let us admit that this regular structure, with generators as points and the similarity group consisting of translations, is only one version, the simplest but certainly not the best possible, of this type of global shape model.

We shall randomize the u, v fields by letting them be solutions to stochastic differential equations

$$\begin{cases} \Delta^p u = e_1 \\ \Delta^p v = e_2 \end{cases}$$

where Δ is the Laplacian operator

$$\Delta = \frac{\partial^2}{\partial x^2} + \frac{\partial^2}{\partial y^2}$$

and e_1, e_2 are Gaussian white noise fields. The notation Δ^p stands for the pth power of Δ. If we introduce the eigen functions $\varphi_{k\ell}(x, y)$ and eigen values $\lambda_{k\ell}$ of Δ

$$\Delta \varphi_{k\ell} = \lambda_{k\ell} \varphi_{k\ell}$$

with appropriate boundary conditions, then u can be written as

$$u(x, y) = \sum_{k,\ell} \lambda_{k\ell}^{-p} z_{k\ell} \varphi_{k\ell}(x, y)$$

where the random variables $z_{k\ell}$ are i.i.d. Gaussian with mean zero (note 6.11). For v we get a similar series expansion. The idea behind this construction is that we want the u, v fields to resemble random elastic deformations of a membrane in directions (longitudinal) of the membrane itself. The full equations for linear elasticity are not quite the ones above, but close enough for the purpose of illustration.

In Figure 9 we show one such deformation. It should be noted that this representation does not guarantee the homeomorphic property. If e_1 and e_2 have large variance we can get maps that fold over, but this is of little practical importance.

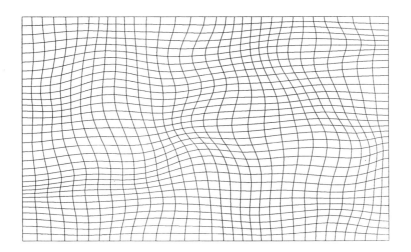

FIGURE 9. HOMEOMORPHIC MAP

6.4. Restoring character strings. In the global shape model we applied group elements to the generators of the template configuration. It is sometimes natural to generalize this by using transformations that do not form a group. We shall illustrate this by carrying out restoration of character strings of LINEAR connection type; these are open patterns and would belong to Chapter 5, but the methodology is that of sections 6.1–6.3.

Consider strings from X^*, where $G = X$ is a finite set that we shall code as $1, 2, \ldots p$, and introduce a probability distribution over X^* as follows. Start with one or several templates, say a single one $c_{temp} = x_1, x_2, \cdots y_n$; $x_i \in X$. For simplicity we shall let $R =$ IDENTITY so that configuration means the same as image. The templates may have been chosen among some regular configuration $\mathcal{C}(\mathcal{R}) \subseteq X^*$ but we shall not go into that here. Instead deform the template configuration by transformations $m \in M = \{m_0, m_1, m_2\}$ with

$$m : G \to \{\phi\} \cup G \cup G \times G$$

meaning that for some $g = x \in X$ we get mg as the empty string, x itself, or the dimer xx. Associate this with positive probabilities p_0, p_1, p_2. The result

$$c = I = \text{LINEAR}(m_1 x_1, m_2, x_2, \ldots m_n x_n) \in X^*$$

is a string whose length ℓ is between 0 and $2n$. All m_i are assumed to be stochastically independent. We shall code m as $\{0, 1, 2\}$.

Given an image $I = \text{LINEAR}(z_1, z_2, \ldots z_\ell)$ we now introduce a noisy version of it. For each i we define

$$y_i = \begin{cases} z_i \text{ with probability } 1 - \epsilon \\ \text{with probability } \epsilon \text{ sample uniformly from } X - \{x\}. \end{cases}$$

The result is the deformed image

$$I^{\mathcal{D}} = \text{LINEAR}(y_1, y_2, \ldots y_\ell).$$

Our task is to restore I by some I^*. To do this we shall *simulate the posterior density of the m_i's given $I^{\mathcal{D}}$*. Once we have the m_i's it is an easy matter to find I. But the m_i's have the prior density, $m = (m_1, m_2, \ldots m_n)$, simply

$$p(m) = \prod_{i=1}^n p_{m_i}.$$

The likelihood can be written

$$L(I^{\mathcal{D}}|I) = \prod_{j=1}^\ell (1 - \epsilon)^{1-e_j} \epsilon^{e_j}$$

where the indicator variables for the errors are

$$e_j = \begin{cases} 1 \text{ if } y_j \neq z_j \\ 0 \text{ else.} \end{cases}$$

Hence the posterior density is proportional to

$$K(m) = \prod_{i=1}^{n} p_{m_i}(1-\epsilon)^{N(m)} \epsilon^{N_e} (1-\epsilon)^{\ell-N_e}$$

where $N_e = \sum_{j=1}^{\ell} e_j$. Note that for fixed y-string $I^\mathcal{D}$ this expression depends only upon m although in a fairly complicated way. If n is big direct simulation of $K(\cdot)$ is computationally awkward or impossible. Instead we shall use a version of stochastic relaxation (see section 6.1).

First initialize m. In principle it does not matter how we do this but convergence will be speeded up by a good choice. The iteration step will be as follows. Select a site $i \in \{1, 2, \ldots n-1\}$, either at random (as will be done here) or by some deterministic sweep strategy. Consider (m_i, m_{i+1}) and the length $\ell_i = m_i + m_{i+1}$. We shall change (m_i, m_{i+1}) into (m'_i, m'_{i+1}) preserving the length $\ell_i = m'_i + m'_{i+1}$ and select the new m'_i, m'_{i+1} by their conditional distribution obtained from $K(\cdot)$. Because of the form of $K(\cdot)$ this will depend only upon that part of the y-string that is deformed from (x_i, x_{i+1}) and is therefore easy to compute.

Then we select another site i, carry out the above modification, and continue iterating a large number of times. Note that there is a finite number of possible m-strings, namely 3^n, and that we can go from any m to any other m' in a finite number of steps with positive probability. Just as was shown in section 6.1 the probability distributions of m will therefore converge to the posterior distribution and the task has been achieved.

Remark. We could not achieve this by modifying a single m_i at a time since $\ell =$ length $(I^\mathcal{D}) = \sum_{i=1}^{n} m_i$ would not be fixed. Therefore we are forced to modify two (or more) m_i's at a time.

In section 9.4 we shall carry out a computer experiment to see how this works out in a special case. At the moment let us just observe that the deformable template idea was here applied when the simple moves (transformations) are not elements in a group. We shall return to this in section 6.8.

6.5. Uncompromising logic. The easiest way to introduce the topic of this section is by an example. Consider the configuration in Figure 10, a simple trimer, and we shall let bonds carry full information, $\beta_j(g) \equiv g, G = \mathbb{R}$, with an acceptor function of the form $A(\beta - \beta')$. Think of the generators g_1 and g_2 as fixed and consider the conditional density of g given g_1 and g_2. It will be, with some constant c,

$$p(g|g_1, g_2) = cA(g - g_1)A(g_2 - g).$$

FIGURE 10. TRIMER CONFIGURATION

It will be convenient to write A in exponential form

$$A(x) = exp[-E(x)]$$

where $E(\cdot)$ is the energy function as in section 4.5. More generally, we would like to make statements about what are likely values in a configuration when only some of the others are observed. One could call such reasoning procedures *uncertain* (not deterministic) *parallel* (not sequential) *logic*, and the above is an extremely simple version.

Assume first that $E(\cdot)$ has its minimum at $x = 0$ and is convex. Then it is clear that

$$p(g|g_1, g_2) \propto exp - [E(g - g_1) + E(g - g_2)]$$

and that because of convexity the function

$$E_{total}(g) = E(g - g_1) + E(g - g_2)$$

will look as in Figure 11(a) with a single minimum at some $g = x_0$. For example, in the Gaussian case $E(g) = g^2$ the minimum is at the average $\frac{1}{2}(g_1 + g_2)$. The most likely value of g, given the observed information, is then $g = g'$, and we will base our decision on this value; or perhaps on a short interval around g'.

In the opposite case, when $E(\cdot)$ is not convex we can get the behavior shown in Figure 11(b). Now we have two minima, g' and g'', and our decision must allow two possibilities. Here our logic does not average, it does not compromise, but offers more than one alternative. The above example is too trivial to bring out fully the importance of this observation: pattern inference should not always produce a single, more or less optimal, result; instead the optimal result can very well be a set of alternatives. In section 9.5 we shall present a more sophisticated version of this principle.

6.6. Inference for deformable templates. We have seen how pattern synthesis based on deformable template representations can be organized; let us now extend this to inference. With notation as in section 6.3 the prior density for the configuration $c = \sigma(g_1, g_2, \ldots g_n) = \sigma(s_1 g_1^0, s_2 g_2^0, \ldots s_n g_n^0)$ can be written as

$$\Pi(c) \propto exp - \frac{1}{2\sigma^2} \sum_{i=1}^{n} \left\{ [u_{i+1} - 1 - a(u_i - 1)]^2 + [v_{i+1} - av_i]^2 \right\}.$$

Plate 1. Sturt, *The Battail of Nasbie*

Plate 2. Vereshchagin, *The Battle of Borodino*

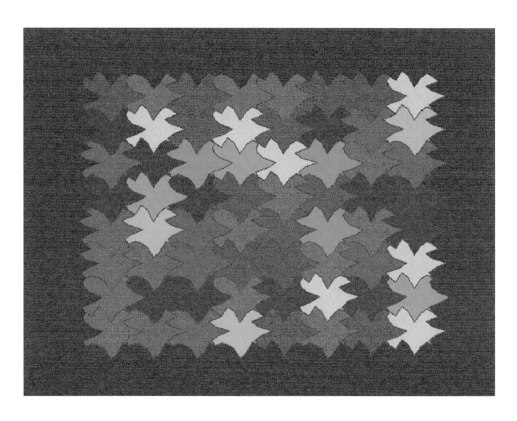

Plate 3. Complete jigsaw puzzle

Plate 4. Incomplete jigsaw puzzle

Plate 5. Textile pattern

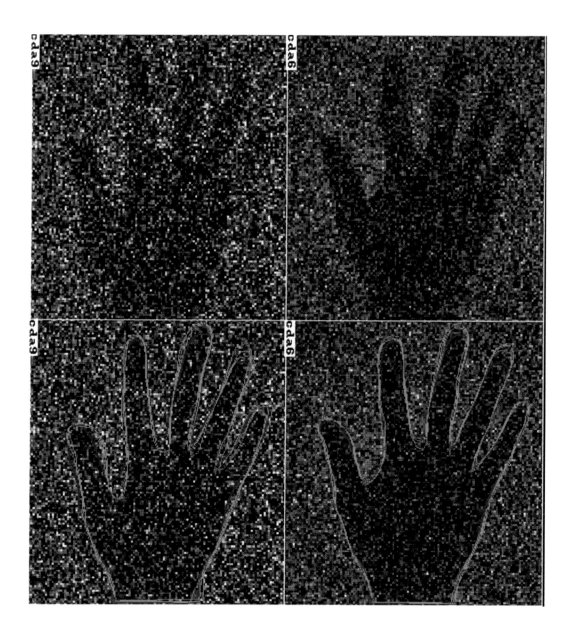

Plate 6. Noisy hand

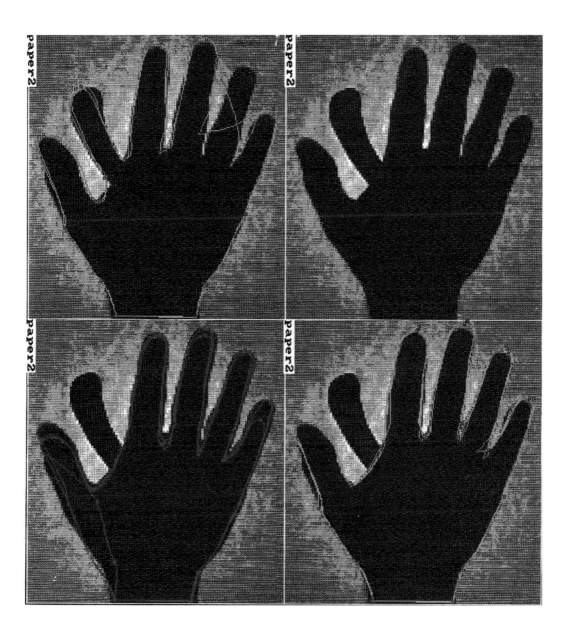

Plate 7. Abnormal hand

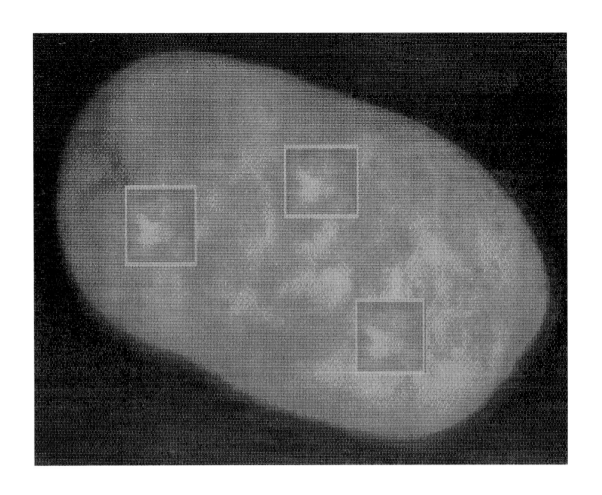

Plate 8. Potato color pattern

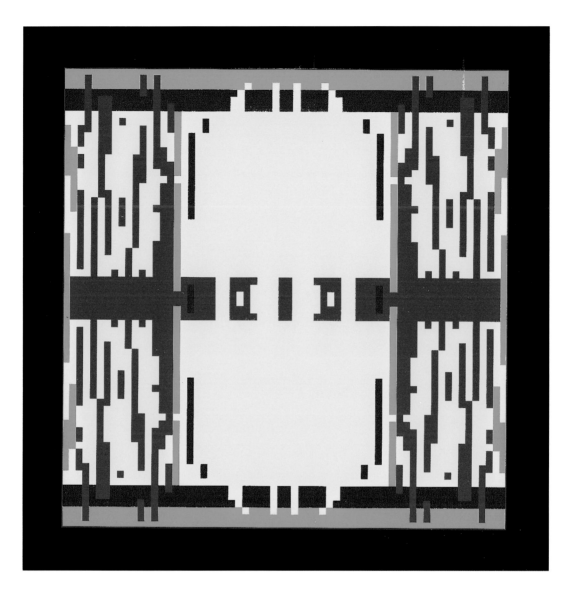

Plate 9. Weave pattern

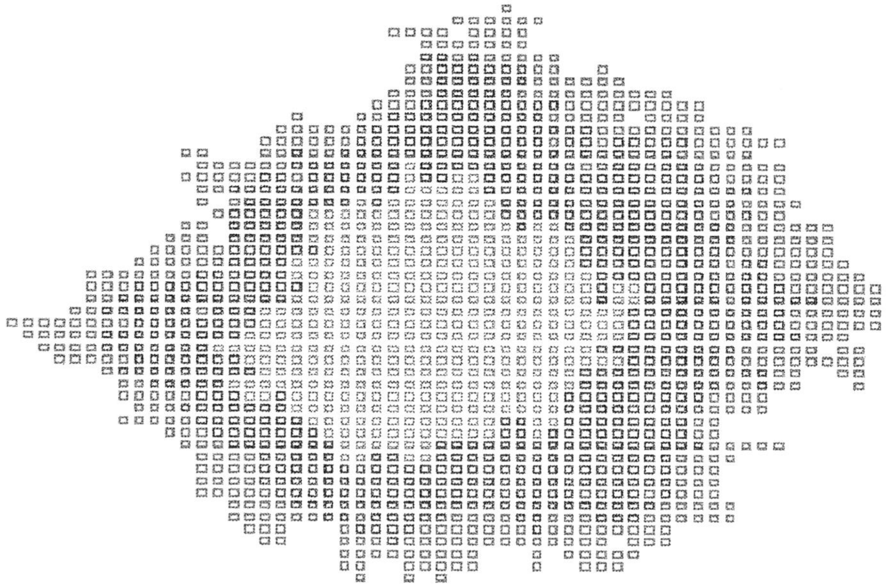

Plate 10. Fairy ring pattern

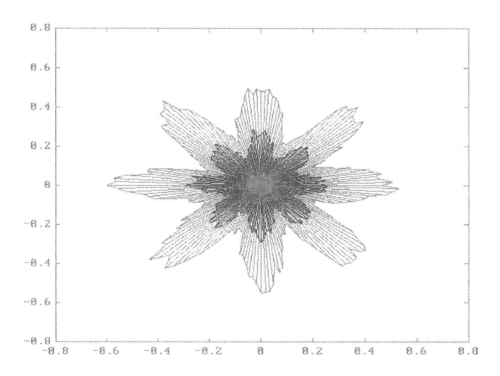

Plate 11. Star-shaped pattern

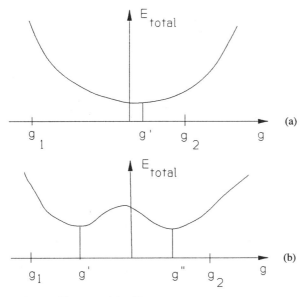

FIGURE 11. ENERGY FUNCTIONS

To derive the likelihood factor (see section 6.1) we must specify the deformation mechanism. The discretized rectangle in which the image is seen will be enumerated $x = 0, 1, \ldots \ell_x$, $y = 0, 1, \ldots \ell_y$ and we shall assume that

$$I^{\mathcal{D}}(x,y) = \begin{cases} m_{in} + e_{xy} & \text{if } (x,y) \in I \\ m_{out} + e_{xy} & \text{if } (x,y) \notin I \end{cases}$$

where I stands for the set inside c and $m_{in} < m_{out}$, and e_{xy} are i.i.d. random variables $N(0, \tau^2)$.

The motivation for this is that in the experiment we are thinking of, HANDS, human hands were placed on a light table with the light source below and the digital camera above the hand. The area inside the hand will therefore have lower light intensity than the outside. To make it harder for the algorithm to achieve good restoration optical degradation was introduced by putting a transparency with non-uniform gray levels between the light and the camera. For example, small particles were glued onto the transparency or cut up paper pieces were attached. Then the likelihood becomes

$$L(I^{\mathcal{D}}|c) \propto exp - \frac{1}{2\tau^2}\left\{ \sum_{inside} [I^{\mathcal{D}}(x,y) - m_{in}]^2 + \sum_{outside} [I^{\mathcal{D}}(x,y) - m_{out}]^2 \right\}$$

where the summations are over $(x,y) \in I$ and $(x,y) \notin I$ respectively. This gives us the posterior density

$$p(c|I^D) \propto exp - \frac{1}{2}\left\{\frac{1}{\sigma^2}\sum_i[u_{i+1} - au_i - (1-a)]^2 + \right.$$
$$+ \frac{1}{\sigma^2}\sum_{i=1}^{n}[v_{i+1} - av_i]^2 + \frac{1}{\tau^2}\sum_{inside}[I^D(x,y) - m_{in}]^2 +$$
$$\left. + \frac{1}{\tau^2}\sum_{outside}[I^D(x,y) - m_{out}]^2 \right\}.$$

We should now apply stochastic relaxation to this; note that varying the s_i's, the u_i, v_i, will affect not only the first two sums but also the last two since I depends upon the s_i's. Our configuration space $\mathcal{C}$ is here a continuum so that the procedure in section 6.1, which assumed $\mathcal{C}$ to be discrete, must be modified. This will not be done here (note 6.12) and we only give two results.

In Plate 6 we display two noisy hand pictures in the top row. The corresponding image restorations are given in the bottom row where green boundary means the true boundary and blue the restored one. The red dots are *landmarks*, distinguished points, at the tip of the fingers and at some other characteristic locations. The result is remarkably good.

In Plate 7, on the other hand, where a false "thumb" has been added, I^D is shown at top right. The other figures show several restored boundaries and only the ones at bottom right look correct. But this is not the right way of evaluating the results. If we instead display the resulting similarity group elements $s_1^*, s_2^*, \ldots s_n^*$ it is seen that some are much too big, namely the ones corresponding to generators between the thumb and the index finger. We have achieved abnormality detection and some image understanding.

A very different application is to images that have internal structure that cannot be described just by a boundary plus textures (say X-ray pictures like Figure 33 in section 3.5). We shall let generators be 3-vectors (x, y, V) where x, y describes location, say in the unit square, and V is a gray-level value. With $\Sigma = $ SQUARE lattice and $S = $ translation group in the plane let us deform the background $(x,y) \to (x + u(x,y), y + v(x,y))$ as in section 6.3. If we observe

$$I^D(x,y) = I(x+y) + e_{xy}; \quad e_{xy} = N(0, \sigma^2),$$

where I is given in section 6.3, we get a likelihood

$$L(I^D|I) \propto exp - \frac{1}{2\sigma^2}\sum_{x,y}[I^D(x,y) - I(x,y)]^2$$

and a simple expression for the prior density

$$\Pi(u, v \text{ fields}) \propto exp - \frac{1}{2\tau^2} \sum_{x,y} \left\{ [\Delta u(x,y)]^2 + [\Delta v(x,y)] \right\}^2.$$

Now we could apply stochastic relaxation to achieve image restoration, which has actually been done, but we shall instead demonstrate another technique that is superior in several respects. Say that we write the posterior density in exponential form

$$p(u, v \text{ fields}|I^\mathcal{D}) \propto \Pi(u, v \text{ fields}) L(I^\mathcal{D}|I) = exp[a(f)]$$

in terms of the fields u and v combined into a field f. Then it is known (note 6.13) that if we solve the stochastic differential equation (S.D.E.)

$$df(t) = grad\ a(f)dt + \sqrt{2}dW(t)$$

where $W(t)$ is the Wiener process with i.i.d. $N(0,1)$ components with the same number of components as $f(t)$, then the probability distribution of $f(t)$ converges to the posterior distribution as $t \to \infty$.

This method has the advantage over stochastic relaxation that it is easier to realize by well-structured computer programs. It uses simpler data structures. It may also converge faster but this is still controversial.

In Figure 12 we show some simulated X-ray pictures of hands. Figure 13 shows image restoration (upper left) of the real image I (lower left) obtained from the noisy picture $I^\mathcal{D}$ (lower right) using the template (upper right). The restoration is not perfect but surprisingly good considering the high noise level in $I^\mathcal{D}$.

6.7. Abnormality detection. Image understanding can be organized in several pattern theoretic modes, one of which deals with automatic detection of abnormalities as was mentioned in the last section. To illustrate how this can be done let us look at a recent study (note 6.4) of inspection of agricultural products, for example potatoes. The object will be characterized by size, shape and color, while position and orientation are considered irrelevant (quotient out the Euclidean similarity group) so that we deal with patterns, not images. Let the template be again $c_{temp} = \text{CYCLIC}(g_1^0, g_2^0, \ldots g_n^0)$ where the g_i^0 are vectors in $\mathbb{R}^2$. We shall let it be deformed into

$$c = \text{CYCLIC}(s^1 s_1^2 s^2 g_1^0, s^1 s_2^2 s^2 g_2^0, \ldots s^1 s_n^2 s^2 g_n^0)$$

where s^1 is a scaling, same for all g_i^0, s^2 is a rotation, also the same, while s_i^2 are rotations varying with i. Of course we ask that the new configuration remains closed.

134 *Chapter 6: Analysis of Closed Patterns*

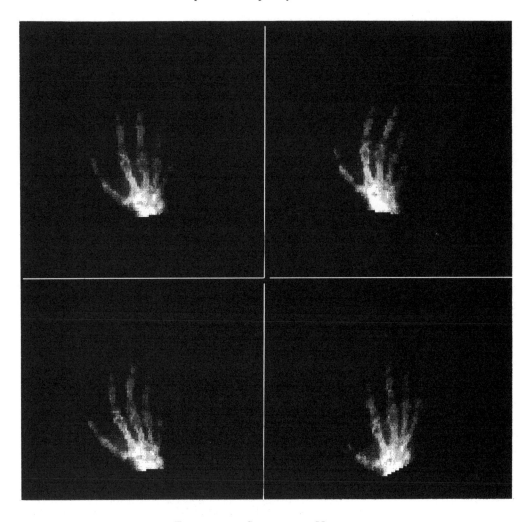

FIGURE 12. SIMULATED X-RAYS

Chapter 6: Analysis of Closed Patterns

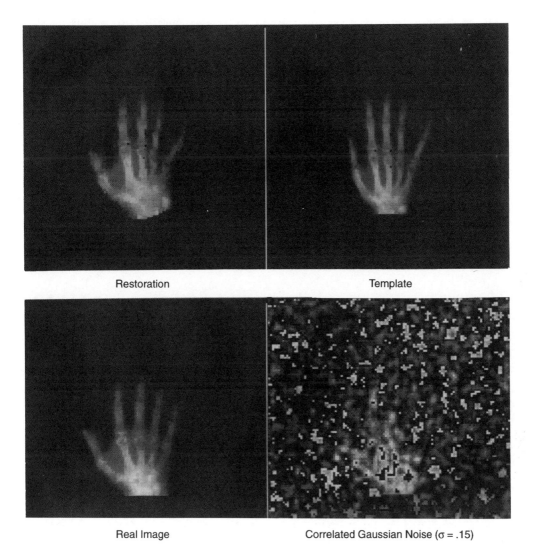

FIGURE 13. X-RAY IMAGE RESTORATION

The probabilities over $\mathcal{C}(\mathcal{R})$ will be introduced by assuming that for scaling $\log s^1 = N(\mu, \tau^2)$, the common rotation angle ϕ of s^2 is uniform over $(0, 2\pi)$ and the rotation angles ϕ_i of s_i^2 we shall assume to have a multivariate von Mises distribution with density

$$p(\phi_1, \phi_2, \cdots \phi_n) \propto exp[b_1 \sum_{i=1}^{n} \cos(\phi_{i+1} - \phi_i) + b_2 \sum_{i=1}^{n} \cos \phi_i].$$

Here b_1 is a coupling parameter and b_2 expresses the amount of deformation; only positive values are of interest. Since we will only have moderate deforming rotations ϕ_i we can approximate, using $\cos x \cong 1 - \frac{x^2}{2}$

$$p(\phi_1, \phi_2, \cdots \phi_n) \propto exp[b_1 n + b_2 n - \frac{b_1}{2} \sum_{i=1}^{n} (\phi_{i+1} + \phi_i)^2$$
$$- \frac{b_2}{2} \sum_{i=1}^{n} \phi_i^2] = exp - \frac{1}{2} Q.$$

Remark. This approximation is not really needed and is only made to simplify the discussion.

The template (of potato shapes) was chosen as an ellipse with semi-axes a and b that were determined from a sample of digital pictures taken inside an imaging chamber. The camera was actually a digital color camera; more about this later on. A picture is shown in Plate 8. The rectangles in it will be explained below. Pattern synthesis was done as has been described earlier with some minor modifications; two configurations are displayed in Figure 14.

Say that we have observed a contour whose successive angles formed by the sides with respect to a fixed direction are ψ_j. We can then write

$$\psi_i = \phi_i^0 + \phi_i + \phi$$

where ϕ_i^0 are the angles for the template, ϕ_i are the rotation angles mentioned and ϕ is the rigid (common) rotation of the whole contour. The log likelihood for $I^\mathcal{D}$ can then be written as

$$\sum_{i=1}^{n} [b_1 \cos(\psi_{i+1} - \psi_i - \phi_{i+1}^0 + \phi_i^0) + b_2 \cos(\psi_i - \phi - \phi_i^0)].$$

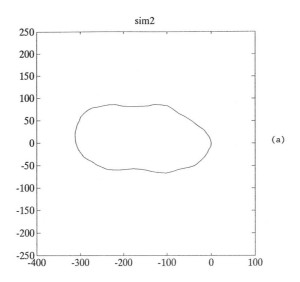

(a)

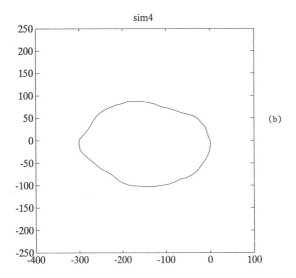

(b)

FIGURE 14. SIMULATED SHAPES

Let us first estimate the rigid rotation angle ϕ that only appears in the second term above. Since

$$\sum_{i=1}^{n} cos(\psi_i - \phi - \phi_i^0) = \cos\phi \sum_{i=1}^{n}(\psi_i - \phi_i^0) + \sin\phi \sum_{i=1}^{n} \sin(\psi_i - \phi_i^0)$$
$$= r\cos\phi\cos\alpha + r\sin\phi\sin\alpha$$

if we introduce

$$\begin{cases} \sum_{i=1}^{n} \cos(\psi_i - \phi_i^0) = r\cos\alpha \\ \sum_{i=1}^{n} \sin(\psi_i - \phi_i^0) = r\sin\alpha. \end{cases}$$

The result is then simply $r\ cos(\phi - \alpha)$ so that the maximum likelihood estimate of ϕ is just α.

We can then write the quadratic form Q in the earlier approximate density as

$$Q = (b_1 t_1 + b_2 t_2 - 2b_1 n - 2b_2 n)$$

with

$$\begin{cases} t_1 = \sum_{i=1}^{n}(\psi_{i+1} - \psi_i - \phi_{i+1}^0 + \phi_i^0)^2 \\ t_2 = \sum_{i=1}^{n}(\psi_i - \alpha - \phi_i^0)^2. \end{cases}$$

Under the assumed approximately Gaussian probability distribution Q will have a χ^2-distribution with n degrees of freedom. Since that is $\sim N(n, 2n)$ for large n we can base the detector for *shape abnormality* on the test

$$q = \frac{Q - n}{\sqrt{2n}} \geq \text{ threshold constant.}$$

In Figure 15 the q-values were for (a) and (b): .36 and 1.8. For (c) and (d), on the other hand, they were 17.1 and 91.2. We then decide the latter two potatoes to have abnormal shape, no doubt due to their sprouts, while the first two are judged normal.

To detect scale (size) abnormality is much easier. Recalling that the template is deformed only by rotations (length preserving) except for s^1, we just have to calculate the length ℓ^0 of the template and ℓ of $I^\mathcal{D}$ and form the criterion

$$r = \frac{\ell/\ell^0 - \mu}{\tau}$$

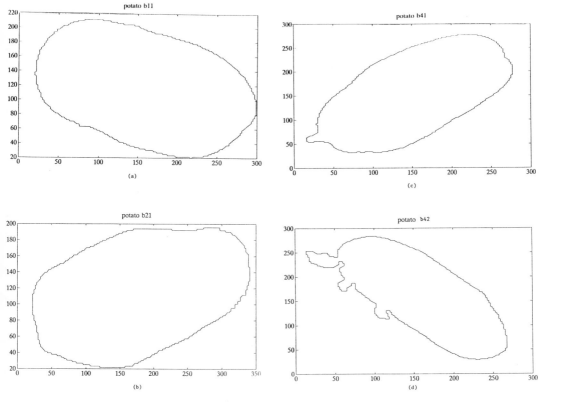

FIGURE 15. POTATO SHAPES

and detect a scale abnormality if $|r|$ is big under $N(0, 1)$ distribution.

A more challenging task is to detect abnormality in the color. For example, a green part of a potato is bad and should be spotted along with other color anomalies. The camera produced three color fields $r(x, y), g(x, y), b(x, y)$ for red, green, and blue. It seemed a natural first step to treat them as nonlinear functions of Gaussian random fields R, G, B of the type discussed in the later part of section 6.3

$$\begin{cases} r(x,y) = f_r[m_r + R(x,y)] \\ g(x,y) = f_g[m_g + G(x,y)] \\ b(x,y) = f_b[m_b + B(x,y)] \end{cases}$$

with $R(\cdot, \cdot)$ satisfying

$$\Delta R + c_r R = \text{white noise} = N(0, \sigma_r^2)$$

and so on. Surprisingly, data analysis of the captured color images showed that the fields were approximately Gaussian themselves so that the f-functions are not needed; they can be just the identity. There must be a physical explanation for this, but it is not known.

A second surprise was that the three error fields e_r, e_g, e_b, that had been guessed to be stochastically independent, were just the opposite. Correlation coefficients of the order .9 occurred especially between red/green, somewhat lower for the other pairs. There must be again a physical, optical, reason for this, but we do not have any convincing explanation for this remarkable fact.

Synthesizing the color fields we get pictures that seem convincingly similar to real (normal) potatoes. To judge the result we have put a small simulated square onto a whole real potato. This is in Plate 8; the result is striking.

After estimating from a sample of normal potatoes the parameters c_r, m_r, σ_r etc., which will not be discussed here, we just form criteria of the type

$$\frac{1}{\sigma_r^2} \sum_{x,y} [(\Delta r)(x,y) + c_2 r(x,y)]^2$$

summed over the inside of a potato image, and detect color abnormality if this criterion is extreme under a χ^2-distribution with appropriate degrees of freedom (normal approximation to χ^2 will do). Applying this to real potatoes, both normal and abnormal, results for the most part were satisfactory (note 6.15).

The above is offered as an example for how pattern theoretic ideas can be used to build automatic detectors of anomalies.

6.8. Multiple objects. So far we have constructed mathematical knowledge representations for images with a single object. But what do we do if we do not know in advance whether there is any object or many objects in the picture? For example, in Figure 44 (section 3.8) we do not know how many mitochondria are present.

To deal with this challenging question we must extend the configuration space. A single mitochondrion could be represented by a configuration space, let us call it $\mathcal{C}_1$, of the type used for the HANDS experiment, with some texture inside and another outside the boundaries. Instead we shall introduce the *multiple object configuration space*

$$\mathcal{C} = \bigcup_{k=0}^{\infty} \mathcal{C}_k = \bigcup_{k=0}^{\infty} \underbrace{\mathcal{C}_1 \times \mathcal{C}_1 \times \cdots \times \mathcal{C}_1}_{k \text{ times}}.$$

Each $\mathcal{C}_1$ can be embedded in $I\!\!R^{2n}$, where $n = $ number of generators for a single boundary, so that $\mathcal{C}$ is in the union (not Cartesian product!) of infinitely many Euclidean spaces.

For fixed k we have to solve, analogously to section 6.6, k S.D.E.'s of the type
$$\begin{cases} dc_\ell(t) = grad\ a[c_\ell(t)] + \sqrt{2}\ dW_\ell(t) \\ \ell = 1, 2, \ldots k. \end{cases}$$
The main difficulty here is the computation of the grad term; we cannot describe the technicalities of how to do that here (see note 6.15).

But we must also let k jump up and down, that is to create or annihilate hypotheses about mitochondria, to carry out abduction. If we start with prior probabilities $\Pi_0, \Pi_1, \ldots \Pi_k, \ldots$ for the probability of having k mitochondria in the picture, the prior can be *written as a sum*
$$\begin{cases} \Pi(c) = \sum_{k=0}^{\infty} \Pi_k p(c_1) p(c_2) \ldots p(c_k) \\ c = \text{union } (c_1, c_2, \ldots c_k) \end{cases}$$
where the connector *union* does not couple the k configurations.

We also have to specify inside/outside textures in order to determine the likelihood. Avoiding some difficult detailed questions, let us just state that the inference machine will operate as follows. Most of the time k will remain fixed and the above S.D.E.'s are solved as time t develops. At random time points, determined by $\Pi(\cdot), L(\cdot|\cdot)$, and $c(t)$ the value of k jumps up one step or down one step. This is illustrated in Figure 16.

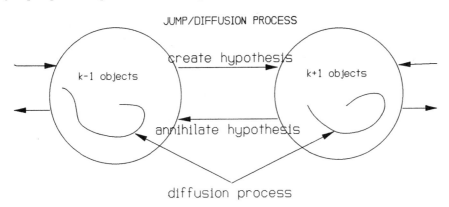

FIGURE 16. INFERENCE MACHINE

To exemplify the working of this inference machine we display some restorations in Figure 17, where the lower right picture is $I^\mathcal{D}$, the observed micrograph in $25,000\times$ magnification. The other three represent different stages of the process with the upper left the last one. It can be seen that the algorithms perform well but make some mistakes, for example take two objects at the left side as a single one. It is believed that most such mistakes can be avoided by making the knowledge representation more detailed.

142   Chapter 6: Analysis of Closed Patterns

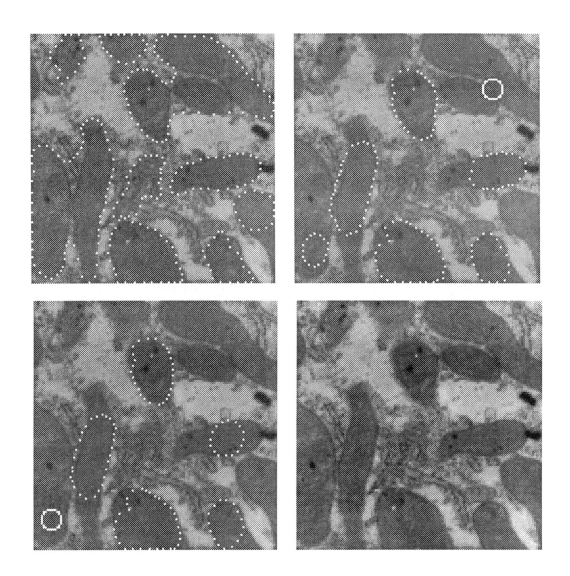

FIGURE 17. MULTIPLE OBJECT RESTORATIONS

PART III

COMPUTING PATTERNS

Chapter 7

Computer Experiments with Patterns

The knowledge representations offered by pattern theory in the form of regular structures are sometimes of unfamiliar type, their functioning not easy to grasp intuitively, and it may be difficult to predict their behavior during the model-building stage: it is sometimes counterintuitive. It is then advisable to experiment with them on the computer, to change parameters and display the results of such changes, to modify the generator space, the acceptor function, etc., in order to get a better insight into how the regular structures behave.

Mathematical experiments on the computer differ from other uses of computers in one important way. In most computing we start with a fairly well-defined task; we want to achieve this task with reasonable speed and expense using the resources available. In the sort of mathematical experiments that we have in mind, however, the task is often nebulous; we do not have a clear idea of what mathematical knowledge representation to use, we will have to try one possibility after another, discarding them if they do not work, modify the code and in general operate in a much less systematic manner.

In such a situation it is imperative that we be able to code and recode quickly so that we can concentrate on the mathematical structures rather than on the programming. Programming is then only a tool, although an indispensable one, our own time is more valuable then computing time, and we do not really aim for well-structured code of great efficiency. On the other hand we want results quickly and displayed in such a way that they help develop our intuition. That usually means graphics, perhaps nothing fancy, but easy to use even at the price of little flexibility.

7.1. Programming language. To plan for experimenting with patterns the first decision to be made is what computing environment we should choose. Since we will be dealing with mathematical objects it is clear that the programming language should make it easy for us to create and manipulate such objects. This seems to imply that we need a high-level language, and this statement is probably correct today. It may change when object-oriented programming becomes available, for example in the form of object-oriented languages like C++ together with suitable mathematical program libraries.

For the time being, however, a high-level language oriented towards mathematics is what we need, and there are many systems available that fit this description. One choice, almost ideal in some respects, is APL (see note 7.1), which was actually developed in order to offer a mathematical symbolism, notation that would incorporate all the basic mathematical structures and operations on them as primitives. Much of the early experimentation in pattern theory was carried out in APL to the great satisfaction of the experimenters.

Once it is learned it is easy to write correct and even elegant APL code quickly, debugging is convenient, and the speed of execution for well-designed APL programs is typically enough for what we need. Its array orientation fits in very well with the regular structures that have been invented, string operations too are adequate, graphics is limited but often sufficient.

But there are drawbacks. It is an unconventional language with its own character set that will mystify the uninitiated and make it awkward to use an ordinary keyboard that has not been prepared for APL characters. It is also harder to learn than the conventional computing languages and, as is often pointed out, difficult to read by other users. More serious than this, though, is the fact that it is not well known or widely used. It is therefore not very suitable for conveying algorithmic information. Somewhat regretfully we have decided not to present the pattern algorithms in APL.

We would like to express the algorithms in a language that can be read without too much trouble by someone who is not very familiar with it: something like pseudo code. At the same time it will certainly help the reader if it can be directly executed too; hence, we have decided to use MATLAB (see note 7.2).

MATLAB can be learned in a few hours, at least the basics of it, it uses ordinary characters, it is fairly well known and available for all the common platforms. The price we pay is that it is very high level with little flexibility. As long as our computing can be expressed in linear algebra terms it is more than adequate; in other situations programming will be more awkward but tolerable. Its graphics and debugging facilities are improving in recent versions; documentation and its HELP facility are well designed.

Although we are presenting the algorithms derived from ideas in Part II in MATLAB form, readers may prefer to recode them into their favorite languages. One possibility is MATHEMATICA (see note 7.3), whose greatest appeal is its elegant conception in mathematically well chosen primitives. It also has excellent graphics including 3-D displays. It is harder to learn than MATLAB, but not to the extent that it presents any serious obstacle. An unusual feature, symbolic algebra, integration, etc., is attractive but seldom needed for our purpose.

Some readers will prefer the more conventional languages like FORTRAN or C. In either case it should be possible to read the MATLAB functions as pseudo code, especially since we will give them together with explanations of what is done in the crucial steps of the programs.

All of this is for preliminary experiments. For large-scale experimentation we get closer to production computing and attention should then be given to speed and memory considerations. But this will not be discussed here since it is outside the scope of this book.

7.2. Hardware. Most readers will carry out the experiments on a PC or Macintosh, but some will use more powerful work stations. The choice of platform is not so important; use whatever is available. It is necessary to be able to display results on the screen and to print them. Color will be useful, and reasonably priced color printers are becoming available (see note 7.4). If the reader has access to such a printer, together with suitable software, it should definitely be used.

Most of the patterns that we will be experimenting with will be synthesized (simulated). We have stressed the importance of pattern synthesis, both for model building and model critique, and for preparing for pattern analysis; the latter can often be organized in a form very close to that of synthesis. Also it is more convenient to synthesize pictures than to capture real ones. Real pictures present additional challenges. Unfortunately they require hardware that many readers will not have. What is needed is, as a minimum, a black/white digital camera together with a monitor and some imaging software for acquiring and manipulating the pictures. Some scanning device could also be used.

Ordinary (visible light) pictures typically include artifacts like shadows, obscuration, and different types of reflections. It requires some technical expertise to handle the resulting problems, but we shall have little to say about this since our main task is knowledge representation of patterns, not image processing technology.

7.3. Programming strategy. To begin a computer experiment, say in pattern synthesis, we first have to define the regularities: generators with bonds, similarities and bond structure groups, acceptor functions and so on. In the simplest cases we can do this directly, for example by defining a bond relation ρ as an $NB \times NB$ matrix, where NB is the cardinality of the bond value space B. When we go to more sophisticated regular structures it can be time consuming and tedious to type in all the parameters in array form. It is then better to use one of the *set-up functions*, if one is available, or else to write one; this will save some effort later on.

Once this has been done we will plan the main program. Although we will not be concerned much with speed it may be necessary to exercise some caution to avoid intolerably long time of execution. Looping should be avoided whenever possible; instead one should exploit the vector/matrix capability. This can result in a dramatic speed-up. Unfortunately, MATLAB does not allow for more than two dimensions (matrices) for the arrays, while many of our synthesis/analysis algorithms are naturally expressed in arrays of three or more dimensions.

Using arrays one can also make the code *appear* parallel; of course looping takes place in the machine but this is not seen explicitly in the program. Such apparent parallelism is of more than superficial interest. Indeed, much pattern computing in recent years has been done employing *parallel computer architecture*, and the Markov structures of many of our knowledge representations are naturally expressed with locality and parallelism.

Once the program has been written and debugged, experiment with it! Change the parameters, pick seemingly absurd values for the parameters, prepare for the unexpected! It is a sad fact that many experiments fail in that they do not do what we want them to do. But sometimes they do something more interesting, something very different from what we expected. And it can be great fun!

It may be that we have set up a generator space that seems well motivated to capture some quality of the patterns that we are dealing with. But the displayed output looks completely wrong. Back to the old drawing board!

Many of our algorithms will be iterative and based on a limit theorem stating some sort of convergence. But theory seldom tells us much about the speed of this convergence, so that if the results look wrong it could be simply because we have not run enough iterations. To get some understanding of this it may be helpful to compute, for each iteration, some statistic about whose asymptotic behavior we have at least some rudimentary knowledge. It is difficult to give precise advice about how to choose such a statistic, perhaps "energy," "temperature," number of closed bonds and so on.

7.4. Some utilities. Let us now develop some simple utility algorithms that will be needed later. A reader may prefer to skip this somewhat pedestrian section and return to it when the utilities are used.

7.4.1. Outer product. Consider two vectors $u = (u_1, u_2, \ldots u_n)$ and $v = (v_1, v_2, \ldots v_m)$. We want to form their outer product in a general sense, that is the $n \times m$ matrix

$$P = (f(u_i, u_j); i = 1, 2, \cdots n, j = 1, 2, \cdots m)$$

where f is a given function of two arguments. To do this we use the MATLAB eval (string) that computes the MATLAB expression given by the character string inside parenthesis.

This is easy:

```
function prod=outer(u,function,v)
%computes the outer product of two vectors u and v
%with respect to a function given by the character
%string 'function' , a valid MATLAB expression
lu=length(u);
lv=length(v);
prod=zeros(lu,lv);
for i=1:lu
        for j=1:lv
                prod(i,j)=eval(['u(i)',function,'v(j)']);
        end
end
```

For example for $u = (1, 2, \ldots 10), v = (0, 1, 2, 3)$ and function meaning "power of," function $=' \wedge '$ we get the outer product

1	1	1	1
1	2	4	8
1	3	9	27
1	4	16	64
1	5	25	125
1	6	36	216
1	7	49	343
1	8	64	512
1	9	81	729
1	10	100	1000

7.4.2. Probability simulation. Let p be a probability vector $(p_1, p_2, \ldots p_n)$ and let us simulate a stochastic variable x that takes the value $x = i$ with probability to p_i. To this end let us first use the MATLAB random number generator to simulate $y = R(0, 1) =$ a uniformly distributed pseudo random number over the range $(0, 1)$. Then compute the partial sums $s_1 = p_1, s_2 = p_1 + p_2, s_3 = p_1 + p_2 + p_3, \ldots s_n = p_1 + p_2 + \ldots + p_n = 1$. If $y > s_1, s_2, \ldots s_i$ but not s_{i+1} we make $x = i + 1$. A moments reflection shows that this does it since the event $s_i < y \le s_{i+1}$ has probability $s_{i+1} - s_i = p_{i+1}$. Hence

```
function x=probsim(probvector)
%simulates a stochastic variable x taking values
%1,2,... with probabilities given by entries in probvector
rand('uniform');
y=rand(1);
x=1+sum(y>cumsum(probvector));
```

If we want x to take values from the vector $v = (v_1, v_2, \ldots)$ instead of $1, 2, 3, \ldots$ we just execute

$$v(\text{probsim}(\text{probvector})).$$

We can also extend the function to produce a whole sample of values instead of a single one.

Another probabilistic simulation utility is for Markov chains. Say we are given a transition probability matrix TRANS, an initial state x and we want to simulate the chain $x_1, x_2, \ldots x_n$ with conditional probabilities

$$P(x_{i+1}|x_i) = \text{TRANS}(x_i, x_{i+1}); i = 1, 2, \ldots n.$$

We can then simply use the utility probsim repeatedly and get the somewhat wasteful program

```
function sample=markov1(init,trans,n)
%simulates Markov chain with initial state init
%transition probability matrix trans
%produces sample of length n
x=init;
sample=[init];
for i=1:n-1
        xnew=probsim(trans(x,:));
        sample=[sample,xnew];
end
```

7.4.3. Displays. We shall need at least some rudimentary displays. For example, if we have generated a matrix $M = (m_{ij})$ with entries in the range $1, 2, \ldots p$ and if we are given a character string "symbols" of length p it is sometimes informative to display the character matrix with entries symbols (m_{ij}). One way of doing this is display 2,

```
function display2(mat,chars)
%displays matrix mat, with entries 1,2,3...p,
%using corresponding entries in character string chars
%of length p
coded=abs(chars);
[m,n]=size(mat);
vec=reshape(mat,m*n,1);
vecchar=coded(vec);
matchar=reshape(vecchar,m,n);
setstr(matchar)
```

where abs(chars) gives the ASCII representation of the character string. In line 7 we reshape mat into a vector and then code it into ASCII, and finally reshape this vector into a matrix and display it as characters using the MATLAB function setstr.

For example if chars $=' B*'$ and mat is

$$\begin{array}{cccc} 2 & 2 & 2 & 2 \\ 2 & 1 & 1 & 2 \\ 2 & 1 & 3 & 2 \\ 2 & 3 & 2 & 2 \end{array}$$

then display2(mat,chars) gives

```
BBBB
B  B
B *B
B*BB
```

Another situation is when we have generated an image represented by a two-column matrix $I = I(k, \ell)$ where $I(\ell, 1)$ and $I(\ell, 2)$ mean the x and y coordinate respectively of the ℓth vertex of a closed polygon. To display it (with closure) we can use the plot function in MATLAB simply by

```
function display1(vertices)
%plots polygon with 'vertices' a 2-column matrix
%the first column x-coordinates,the second y-coordinates
%with closure
xs=[vertices(:,1)',vertices(1,1)];
ys=[vertices(:,2)',vertices(1,2)];
plot(xs,ys)
```

but note lines 5,6 for the closure condition.

If we define the columns in "vertices" by

$$\begin{cases} \text{vertices}(i,1) = \cos 4\frac{2\pi i}{n} \\ \text{vertices}(i,2) = \sin 3\frac{2\pi i}{n} \end{cases}$$

and apply display1 (vertices) we get the Lissajou curve in Figure 1.

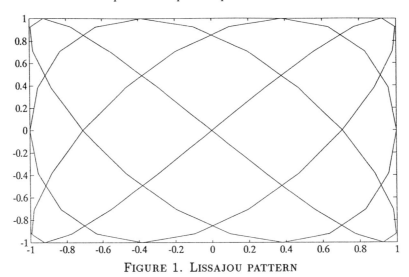

FIGURE 1. LISSAJOU PATTERN

We shall also sometimes display single points stored in a two-column matrix here called id:

```
function seeid1(id)
%displays point scatter in 2-column matrix id=deformed image
clg
l=length(id(:,1));
axis([-1 1 -1 1])
hold on
for k=1:l
        plot(id(k,1),id(k,2),'*')
end
```

The reader may prefer to replace the character * by some other.

7.4.4. Acceptor matrices. To compute products of acceptor matrix values

$$A(z_1, z_2)A(z_2, z_3) \ldots A(z_{m-1}, z_n)$$

we use the auxiliary self-explanatory program

```
function product=accept1(zs,A)
%computes product of values from acceptor matrix A
%over configuration string zs
product=1;
m=length(zs);
for i=1:m-1
        product=product*A(zs(i),zs(i+1));
end
```

7.4.5. Selection.
To select entries from a given vector can be done in many ways, one of which is

```
function result=select(vector,boole)
%selects those entries in the vector where
%the corresponding entry in boole is a 1
%boole should be a boolean vector , with zeros and ones
%of the same length as vector
n=length(vector);
result=[];
for i=1:n
        if boole(i)==1;
        result=[result,vector(i)];
        end

end
```

Chapter 8

Computing Open Patterns

We are now ready to experiment with regular structures and start with those of the simplest graph structure: the open patterns as described in Chapter 5. Since the connectors then will have no cycles there is no need for iterative procedures with all the trouble that entails. Instead we can employ direct algorithms and we need not worry about the speed of convergence.

8.1. Character strings. To synthesize configurations like the ones in section 5.1 with relaxed regularity is easy since their probability distributions are given by Markov chains. It takes a bit more effort to program the abnormality detector. Note however that the long products of acceptor values for p_0 and p_i can be evaluated by the utility accept1 in section 7.4. The matrix power $A^{\ell+1}$ can be done in MATLAB primitives: $A \wedge (\ell + 1)$.

With these observations we can express the p_i's and q_0 in section 5.1 as in abnorm1:

```
function [q0,ps]=abnorm1(config,A,l,pabnorm)
%automatic abnormality detector for configurations
%of character strings
%A is acceptor matrix, say size p*p
%l is length of possible abnormal sub-configuration
%pabnorm is prior probability of abnormality occurring
%config should be string with values 1,2...p
%q0 is posterior probability of abnormality
%ps is vector of values proportional to probability of abnormality
%beginning at sites;only 2,3...n-1 are meaningful
Apower=A^(l+1);
p=length(A(:,1));
n=length(config);
p0=(1-pabnorm)*accept1(config,A);
ps=zeros(1,n-1);
for i=2:n-1
        ps(1,i)=(pabnorm/(n-1-1))*accept1(config(1:i-1),A)*Apower(config(i-1),c\
onfig(i+1))*accept1(config(i+1+1:n),A);
ps(1,i)=ps(1,i)*(p^(-1));
end
q0=sum(ps(2:n-1))/(p0+sum(ps(2:n-1)));
```

It should be observed, however, that if n is very large the long products obtained by accept1 better be done in terms of sums of logarithms in order to avoid over/under flow; this requires that all entries in A are strictly positive.

We can now experiment with the abnormality detector. With $n = 100$, $p_{abnorm} = 20\%$, $\ell = 10, p = 5$, and

$$A = \begin{matrix} 0.8000 & 0.0500 & 0.0500 & 0.0500 & 0.0500 \\ 0.0500 & 0.8000 & 0.0500 & 0.0500 & 0.0500 \\ 0.0500 & 0.0500 & 0.8000 & 0.0500 & 0.0500 \\ 0.0500 & 0.0500 & 0.0500 & 0.8000 & 0.0500 \\ 0.0500 & 0.0500 & 0.0500 & 0.0500 & 0.8000 \end{matrix}$$

and using the alphabet alph= [$'abcde'$] we get a normal configuration confign and print alph(confign)

aaaaaaaaaaaaaabbbbbbbbbbbbbbbcccdddddddeeeeeeeeeeeeaabbbbbbbbbbbbbbbbbbbbbbbbb bbbbbcccccccdeeeeeeeee

Applying abnorm1 to this string we get the posterior probability of abnormality as $5.8012 \cdot 10^{-4}$.

If we replace a substring of confign of length 10 by a purely random one (use markov1 with trans = .2*ones(5)), we get an abnormal string confign, printed as alph(confign)

aaaaaaaaaaaaaabbbbbbbbbbbbbbbcccdddddddeceecacebcbeeabbbbbbbbbbbbbbbbbbbbbbbb bbbbbccccccdeeeeeeeee

and the posterior abnormality probability jumps to .9997.

The reader may want to extend this to cases where the length of potentially abnormal substrings are not fixed but random, or where the abnormal substring is not purely random (with stochastically independent characters) but is generated as a Markov chain with some other acceptor matrix given as trans in markov1. The case that we have studied is admittedly a simple one, but it illustrates the general approach to abnormality detection and we shall have more to say about it applied to more sophisticated regular structures.

8.2. FS and CF patterns. To synthesize the language patterns from Chapter 5 we first need some setup functions for defining the dictionary of words and the generator space with attached weights inducing probabilities.

We begin with the dictionary. If the longest word of length is called maxlength and if we employ nwords different words we shall create two arrays. One, called words, is going to be a matrix with nwords rows and maxlength +1 columns. We have added an extra column since we want a word like "boy" with a blank attached at the end, which will be convenient to separate words in the sentence.

The second array is a vector, called wlengths, that tells us how long a word is (with a blank attached). These two arrays are computed by setupwords:

```
function [words,wlengths]=setupwords(nwords,maxlength)
%helps define dictionary with nwords items
%the longest one of length maxlength
words=[];
wlengths=[];
maxlength=maxlength+1;
space=' ';
for i=1:nwords
        sprintf('word no. %g',i)
        word=input('enter word:','s')
        l=length(word);
        newword=[word ,space(ones(maxlength-1,1))];
        words=[words;newword];
        wlengths(i)=l+1;
end
```

Note in line 12 how we attach a number of blanks to make the word fit into a row vector of fixed length.

We also set up the generator space, call gspace, by the function setupfs:

```
function gspace=setupfs(ngen)
%helps define finite state grammar with ng
%generators (rewriting rules)
gspace=[];
for i=1:ngen
        sprintf('generator no. %g',i)
        vec=input('ixj-vector and probability=');
        gspace=[gspace;vec];
end
```

The user inputs vectors $[i\ x\ j\ p]$ where the generator is $i \xrightarrow{x} j$ and p is the attached weight to define probabilities; see line 7.

With these utilities it is easy to write the synthesis program

```
function image=synthfs(gspace,words,wlengths)
%synthesizes finite state language images
%needs generator space gspace
%dictionary words
%and vector of wordlength
ngen=length(gspace(:,1));
image=[];
state=1;
final=max(gspace(:,3));
while state<final
        present=state==gspace(:,1);
        next=select([1:ngen],present');
        probs=gspace(next,4);
        to=next(probsim(probs));
        state=gspace(to,3);
        image=[image,words(gspace(to,2),[1:wlengths(gspace(to,2))])];
end
```

In this program we initialize the state $i = 1$, we compute the largest state value final, by convention this will be the final state. Using the utilities select and probsim from section 8.1 we then simulate the Markov chain, jumping from one state to another with probabilities given by the vector probs. The only tricky part is in the next to the last line where we choose the right number of characters from words for the particular generator used. To illustrate this let us think of a language fragment that might appear in one of the folk tales discussed in Chapter 2. The dictionary words will be chosen as

the	4
young	6
ugly	5
pretty	7
evil	5
prince	7
boy	4
princess	9
girl	5
who	4
lived	6
owned	6
in	3
some	5
big	4
small	6
cottage	8
house	6
castle	7
walked	7
went	5
to	3
a	2
big	4
dark	5
deep	5
forest	7
lake	5
.	2
and	4

with the corresponding vector wlengths beside it.

The generator space will be

gspace =

1.0000	1.0000	2.0000	1.0000
2.0000	2.0000	2.0000	0.1250
2.0000	3.0000	2.0000	0.1250
2.0000	4.0000	2.0000	0.1250
2.0000	5.0000	2.0000	0.1250
2.0000	6.0000	3.0000	0.1250
2.0000	7.0000	3.0000	0.1250
2.0000	8.0000	3.0000	0.1250
2.0000	9.0000	3.0000	0.1250
3.0000	10.0000	4.0000	0.2000
3.0000	20.0000	8.0000	0.4000
3.0000	21.0000	8.0000	0.4000
4.0000	11.0000	5.0000	0.5000
4.0000	12.0000	6.0000	0.5000
5.0000	13.0000	6.0000	1.0000
6.0000	14.0000	7.0000	0.5000
6.0000	1.0000	7.0000	0.5000
7.0000	15.0000	7.0000	0.2000
7.0000	16.0000	7.0000	0.2000
7.0000	17.0000	3.0000	0.2000
7.0000	18.0000	3.0000	0.2000
7.0000	18.0000	3.0000	0.2000
3.0000	20.0000	8.0000	0.5000
3.0000	21.0000	8.0000	0.5000
8.0000	22.0000	9.0000	1.0000
9.0000	1.0000	10.0000	0.5000
9.0000	23.0000	10.0000	0.5000
10.0000	24.0000	10.0000	0.1000
10.0000	25.0000	10.0000	0.1000
10.0000	26.0000	10.0000	0.1000
10.0000	27.0000	11.0000	0.3000
10.0000	28.0000	11.0000	0.4000
11.0000	29.0000	12.0000	0.8000
11.0000	30.0000	1.0000	0.2000

which is not very informative. A more intuitive way of visualizing it is by Figure 1, which gives the wiring diagram of the automaton that generates the regular structure of sentences.

160 Chapter 8: Computing Open Patterns

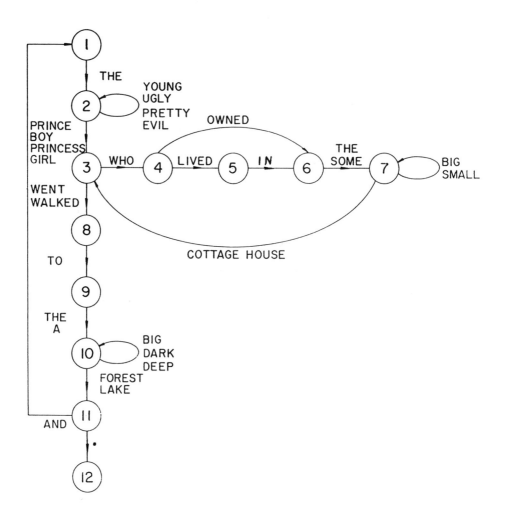

FIGURE 1. FINITE STATE AUTOMATON

Running the automaton synthfs(gspace,words,wlengths) we get reasonable-looking sentences like

the boy who owned the small cottage went to the deep forest.

the prince walked to the lake.

the girl walked to the lake and the princess went to the lake.

the pretty prince walked to the dark forest.

and occasionally more awkward ones

the evil evil prince walked to the lake.

the prince walked to the dark forest and the prince walked to a forest and the princess who lived in some big small big cottage who owned the small big small house went to a forest.

The reader should experiment with the synthesis by allowing $A(\cdot, \cdot)$ to take other values than 0 and 1 so that common grammatical mistakes can occur occasionally. For example, double negations (**I do not see nothing**) or disagreement in number (**they is not here**). Indeed, colloquial language does not follow rigid grammatical laws; its patterns are more flexible. To allow $A(\cdot, \cdot)$ to take other values than 0 and 1 is a regularity-breaking device that represents higher variability. This remark is valid in greater generality, for example in medicine, when rigid regularity is not enough for a realistic knowledge representation. The reader should speculate about such possibilities!

To prepare for synthesizing context-free strings we have to set up the generator space differently. Remembering that a generator here means a rewriting rule, say of the form

$$x \to x_1 \ x_2 \ x_3 \ldots x_\ell$$

where $x \in V_N$ and the other x_i's are in $V = V_N \cup V_T$ we shall let gspace be a matrix of size $ng * 10$. Here ng is the number of generators and we assume (arbitrarily) that the length of the full rewriting rule $1 + \ell \leq 10$.

Let us also use the convention that elements in V_N are enumerated by $101, 102, 103, \ldots$ while numbers smaller than 100 are reserved for elements in V_T. Then setupcf interrogates the user about successive generators and their probabilities:

```
function [gspace,ps]=setupcf(ngen)
%prepares for using synthcf
%ngen number of generators
%input [x x1 x2 ...] where x is number of non-terminal to be rewritten
%and x1,x2... is rewritten string
%also input corresponding probability
%author Peter Schay,1992
gspace=[];
for i=1:ngen
        sprintf('generator no. %g',i)
        vec=input('enter [x x1 x2...]');
        p=input('enter probability');
        ps=[ps;p];
        vec=[vec,zeros(1,10-length(vec))];
        gspace=[gspace;vec];
end
```

To synthesize sentences we shall start with the first variable in V_N coded 101. We build up a derivation tree, looking for non-terminals as long as there are any, if not the derivation ends. Finding a non-terminal x we look in the first column in gspace for occurrences of x and their probabilities. Simulate the probabilities to find one of these generators and join the subtree with the upper node x to the already existing tree. This is done by synthcf (note 8.1),

```
function sentence=synthcf(gspace,ngen,words,wlengths,ps)
%synthesize sentences from context free language
%with generator space given as matrix with ngen rows
%produced by function setupcf as well as probability vector ps
%and words,wlengths produced by setupwords
%author Peter Schay,1992
sent=[101];
done=0;
while (done==0) % done is 1 when no non-terminals left
done=1;
newsent=[];
for i=1:length(sent);
        if(sent(i)>100), % if it is a non-terminal
           done=0;
%find one of the rewriting rules for this non-terminal
        boole=sent(i)==gspace(:,1);
        rows=select([1:ngen],boole');
        probs=ps(rows);
        row=rows(probsim(probs));
        rowlen=sum(gspace(row,1:10)>0);
%copy the rule to the new sequence
        newsent=[newsent,gspace(row,2:rowlen)];
        else
        newsent=[newsent,sent(i)]; %copy the terminal to newsent
        end
end
sent=newsent ;% repeat the process
end
for i=1:length(sent)
        sentence=[sentence,words(sent(i),1:wlengths(sent(i)))];
end
```

Some explanations may be in order. The variable *done* means completion of derivation when *done* = 1. The string of leaves of the current string is called *sent*, and in line 13 we look for non-terminals. If we find one, line 16 locates it in the first column of *gspace* and also gets the associated probabilities from the vector *ps*. They are simulated using the utility probsim from section 7.4 and we build a new string *newsent* in line 22 using the entries in column 2, 3, ... of *gspace* for the row meaning the non-terminal found. We then continue until no more non-terminals are found.

Let us apply this to algebraic expression and introduce the dictionary and non-terminals

words =	wlength
(	1
)	1
sqrt	4
+	1
−	1
*	1
/	1
^	1
sin	3
cos	3
log	3
x	1
y	1
z	1

non-terminals =
exp
op
num
digit

Here we have made the entries in wlength (see above) one less than before, since we do not want to separate words by blanks.

The rewriting rules are chosen as

exp → (exp) op (exp)	exp → cos(exp)
exp → num	exp → sin(exp)
exp → sqrt(exp)	exp → log(exp)
op → +	num → digit
op → −	num → num digit
op → *	digit → x
op → /	digit → y
	digit → z

Applying setupcf (16) we get gspace

```
gspace =
        101    1   101    2   102    1   101    2    0    0
        101  103     0    0     0    0     0    0    0    0
        101    3     1  101     2    0     0    0    0    0
        102    4     0    0     0    0     0    0    0    0
        102    5     0    0     0    0     0    0    0    0
        102    6     0    0     0    0     0    0    0    0
        102    7     0    0     0    0     0    0    0    0
        102    8     0    0     0    0     0    0    0    0
        101    9     1  101     2    0     0    0    0    0
        101   10     1  101     2    0     0    0    0    0
        101   11     1  101     2    0     0    0    0    0
        103  104     0    0     0    0     0    0    0    0
        103  103   104  104     0    0     0    0    0    0
        104   12     0    0     0    0     0    0    0    0
        104   13     0    0     0    0     0    0    0    0
        104   14     0    0     0    0     0    0    0    0
```

Now execute synthcf(gspace,16,words,wlengths,ps) which produces algebraic expressions like

$$\log(x)$$
$$\sin(\cos(\cos(y)))$$
$$\log(\log(z))$$
$$\sin(sqrt(sqrt(\log((y) + (z))))).$$

Sometimes the expressions will be more complicated:

$(\log(\sin(sqrt(\sin(\log(sqrt((\cos(\cos(\sin(sqrt(sqrt(\sin(\log(\log(z))))))))) - (x))))))))$
$* (\sin(\sin(\log(sqrt(\log(sqrt(\sin(y))))))))$

or the awesome expression

```
log((sqrt(sin(log(log((z)-((sqrt(log(cos(log(sqrt(cos(log(sqrt((log(log(sin((sqr
t(sin(sqrt(cos(sqrt(log(((sin(sqrt(sqrt((x)*((log(sqrt(sin(sqrt(log(log(log((zzx
)+(sin(sin(cos(sin(sqrt(log(z))))))))))))))))+(cos(cos(cos(x))))))))))^(sqrt(cos(s
qrt(sqrt(sin((sin(sin(cos(yyx))))-(log(z))))))))))*(sin(sqrt((sqrt(sqrt(sin(x))))
*((y)/(log(log(y))))))))))))))-(y)))))/(sqrt(y))))))))))))+(z))))))^(sin(((sqrt(
log(sin(cos(sin(cos(log(sin(cos(cos(sqrt(z))))))))))))))^(sqrt(log(sin(sqrt(z)))))
)-(sin(log((cos(log(cos(sqrt(sin(sqrt(z))))))))/(y))))))))
```

The probability of getting such unwieldy expressions is controlled by the entries in the vector ps as was discussed in section 5.4. These expressions are purely syntactic in that no attempt has been made to remove unnecessary parentheses or simplify the expression by the usual algebraic rules. To do that is a very different task faced by computer scientists who build programs for algebraic symbol manipulation, integration, differentiation, and so on.

Context-free grammars can be used to represent many formal systems such as programming languages. They do not approximate the intricate structure of natural language well, however, although they do this better than finite state grammars. Human languages are biological systems of high complexity and potential beauty and it would be scientifically naive to believe that they can be represented accurately and completely by simple logical constructs. These patterns look complex and *are* complex! The very notion of "syntactically correct" is suspect with its limiting dichotomy right/wrong. Something more flexible is needed, perhaps some random mechanism that affects both the generators and the connectors. But how?

8.3. Computing regime patterns. The waveform patterns in section 5.2 with changing regimes driven by a character string configuration requires a numerical differential equation solver. We shall use MATLAB's function ode23 which has the format

$$[ts, xs] = ode23('name', t0, t1, init)$$

where $t0$ is the initial time point, $t1$ the final one, and init the initial vector value. "Name" is the name of a function that should have the form in our case, where the differential operators are of order 2:

$$\text{function } xdot = \text{funct}(t, x)$$
$$xdot(1) = \text{expression in } x(1), x(2)$$
$$xdot(2) = x(1).$$

The waveform synthesizer will take as input a string of numbers from $1, 2, \ldots p+1$ called config1, and a number lengths that denotes the length ℓ in section 5.2. The output $[ts, xs]$ consists of $ts =$ vector of time points, and $xs =$ a two-column matrix whose second column means the resulting values. To display the waveform we do

$$\mathrm{plot}(ts, xs(:, 2)).$$

The code could look like this:

```
function [ts,xs]=synthwaves1(config1,lengths)
%synthesizes wave form images
%config1 is configuration of generators
%meaning operators called grad1,...
n=length(config1);
xs=[.5 .5];
ts=0;
for i=1:n
        last=length(xs(:,1));
        [newts,newxs]=ode23(grads(config1(i),:),lengths*(i-1),lengths*i,xs(last\
,:));
        ts=[ts;newts];
        xs=[xs;newxs];
end
```

The computer experiment referred to in Chapter 5 used the following five functions, all obtained as gradients of energy functions:

```
function xdot=grad0(t,x)          function xdot=grad1(t,x)
xdot(1)=-x(2)^3;                  xdot(1)=4*x(2)*(1-x(2)^2);
xdot(2)=x(1);                     xdot(2)=x(1);
```

```
function xdot=grad2(t,x)          function xdot=grad3(t,x)
xdot(1)=-x(2);                    xdot(1)=-2*x(2)*exp(-x(2)^2);
xdot(2)=x(1);                     xdot(2)=x(1);
```

```
function xdot=grad4(t,x)
xdot(1)=-.3*x(2);
xdot(2)=x(1);
```

The function synthwaves1 calls a character matrix grads:

$$grad0$$
$$grad1$$
$$grad2$$
$$grad3$$
$$grad4$$

which should be made global. The function grad1 computes the gradient of the function $(1-x^2)^2$ (and changes the sign) which has *two* minima for $x = \pm 1$. Manipulating constants in the grad functions one can get fascinating regime behavior extending what we saw in section 5.2. This should be tried.

8.4. Computing curve images.
The computer implementation of the deformed curve templates in section 5.5 is not difficult. Let us do it for an example that starts from a ray template and deforms it by a von Mises distribution over the orthogonal group 0(2) of rotations in the plane.

We need to simulate the conditional von Mises densities proportional to, for fixed e',

$$exp[a\ cos(e'' - e') + b\ cos(e'')]$$

and do this by brute force. Discretize the range $[0, 2\pi)$ of the angle e'' into a set of n discrete values equi-distantly spaced. Then we compute the above expression for e'' taking each of these values, normalize to probabilities by dividing by their sum. This done we can just call the utility function probsim given in 7.4 and iterate for $i = 1, 2, \ldots n =$ number of generators (vectors):

```
function angles=vonmises(n,ndiscrete,a,b)
%simulates multivariate von Mises distribution
%sample size=n;ndisrete number of allowed angles
%%coupling parameter=a, pekedness parameter=b
angle=0;
angles=[];
discangles=(2*pi/ndiscrete).*[0:ndiscrete-1];
for i=1:n
        probs=exp(a.*cos(discangles-angle)+b.*cos(discangles));
        probs=probs./sum(probs);
        x=probsim(probs);
        angle=discangles(x);
        angles=[angles,angle];
end
```

To deform the ray template, all of whose generators are unit vectors with orientation angles $\varphi_i + e_i$ (see section 5.5 for notation), we treat each ray separately (!), so that we just have to compute the e_i's from the function von Mises, rotate the unit vectors from orientation φ_i (uniformly spaced) by adding the e_i's. Then we form the vector sums as cumulative sums of the $x-$ and $y-$components and plot the new curves:

```
function deftempl(nradii,n,ndiscrete,a,b)
%synthesizes and displays image of rays
%using deformable template with nraddii rays
%and n generators along each ray
%ndiscrete=number of discretized angles
axis([-n,n,-n,n]);
for k=1:nradii
        angles=vonmises(n,ndiscrete,a,b);
        defangles=angles+2*k*pi/nradii;
        xs=cos(defangles);
        xs=cumsum(xs);
        ys=sin(defangles);
        ys=cumsum(ys);
        plot(xs,ys);
end
```

A reader may wish to write a more general program that deforms a given curve template in general by some other deformation than von Mises. One possibility worth experimenting with is when the template generators are deformed by

$$g_i^0 \to A_i g_i, \quad A \text{ random } 2 \times 2 \text{ matrix}$$

is to choose similarities as the 2×2 matrices

$$\begin{pmatrix} 1 + a_{i11} & a_{i12} \\ a_{i21} & 1 + a_{i22} \end{pmatrix}$$

where all a's are Gaussian and of Markov type

$$a_{ik\ell} = b a_{i-1 k\ell} + e_{ik\ell}$$

with all $e_{ik\ell}$ i.i.d. $N(0, \sigma^2)$ and b is a positive coupling constant. Here $S_2 = GL(2)$, the general linear group in two dimensions (note 8.2).

8.5. Snake shapes. Let us choose as generators line segments of fixed length (called dl in the code) and a template as exemplified in Figure 2. The similarity group S will be selected as consisting of transformations

$$\begin{cases} x_i^0 \to x_i = x_i^0 + a_1 + a_2 i/n + a_3 (i/n)^2 \\ y_i^0 \to y_i = y_i^0 + a_4 + a_5 i/n + a_6 (i/n)^2 \end{cases}$$

where (x_i^0, y_i^0) are the vertices of the polygonal template and (x_i, y_i) those of the resulting configuration. The dimension of S is here 6.

The probability distribution of the a_i's will be assumed to be $N(0, \sigma^2)$, all a_i's i.i.d. We have done this to illustrate the fact that we do not really have to assume Markovian nature for the distributions on $\mathcal{C}(\mathcal{R})$, although we have usually done this. A realization is shown in Figure 3; it has produced a major change, the curve intersects itself.

The template has been produced by the following function, which is of limited interest but is adequate to illustrate the methodology:

```
function [xs,ys]=snake1(npoints,dl,period)
%computes snake template with npoints points
%and line segments of length dl as generators
fis=(2*pi/period).*[0:npoints-1];
fis=fis.^2;
fis=fis.*.1;
xs=dl.*cos(fis);
ys=dl.*sin(fis);
xs=.2+cumsum(xs);
ys=.5+cumsum(ys);
```

Chapter 8: Computing Open Patterns

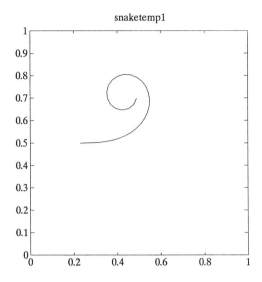

FIGURE 2. SNAKE TEMPLATE

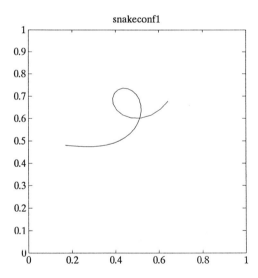

FIGURE 3. SNAKE CONFIGURATION

The resulting configuration obtained by deforming the template is also quite straightforward, and an example is shown in Figure 3,

```
function [cxs,cys]=snakedefl(xs,ys,sigma)
%deforms snake with coords xs and ys into config
%with coords cxs and cys
% and scaling sigma
rand('normal')
npoints=length(xs);
v=[0:npoints-1]./npoints;
random=sigma.*rand(1,6);
cxs=xs+random(1,1)+v.*random(1,2)+random(1,3).*v.^2;
cys=ys+random(1,4)+random(1,5).*v+random(1,6).*v.^2;
```

The deformation mechanics $\mathcal{D}$ is, however, of considerable interest. It will be written $\mathcal{D} = \mathcal{D}_2 \mathcal{D}_1$, when $\mathcal{D}_2$ is just additive noise (here with standard deviation .2), $\mathcal{D}_1$ is a *spreading* or widening transformation. The configuration with vertices (x_i, y_i) will give rise to a deformed digital image id, say a matrix of size npoints*npoints with intensities $id(x, y)$ given by a *point spread function* of Gaussian form

$$id(x,y) = \sum_{i=1}^{n} exp(-c(x-x_i)^2 - c(y-y_i)^2).$$

The coefficient c is inversely proportional to the widening effect. This mechanism is believed to realistically represent some actually occurring deformations, for example in handwriting. This is coded into the function snakespread1:

```
function id=snakespread1(cxs,cys,nimage, c)
%computes deformed digital image id of size nimage*nimage
%for snake config. with coords. cxs and cys
% c inverse of length of spreading
n=length(cxs);
id=zeros(nimage,nimage);
v=[1:nimage]./nimage;
for i=1:n
        dist1=(v-cxs(i)).^2;
        dist2=(v-cys(i)).^2;
        dist1=exp(-c.*dist1);
        dist2=exp(-c.*dist2);
        mat=kron(dist1,dist2');
        id=id+mat;
```

For $\mathcal{D}_2$ we simply add Gaussian noise $N(0, \tau^2)$ and obtain $I^\mathcal{D}$. The question is, how can we apply the principles in Part II to construct an image restoration algorithm? The potential energy of the posterior will come from the prior, simply,

$$E_{prior} = \frac{1}{2\sigma^2} \sum_{i=1}^{n} a_i^2$$

and from the likelihood, more complicated,

$$E_{likely} = \frac{1}{2\tau^2} \sum_{x,y} \{I^\mathcal{D}(x,y) - \sum_{i=1}^n p[(x,y),(x_i,y_i)]\}^2$$

summed over all x,y-values of the digital image $I^\mathcal{D}$ and where

$$p[(x,y),(x',y')] = exp\{-c[(x-x')^2 + (y-y')^2]\}.$$

The (x_i, y_i) are the points on the template.

We therefore get the gradient of $E = E_{prior} + E_{likely}$ as

$$\frac{1}{\sigma^2} a_k - c \sum_{x,y} \{I^\mathcal{D}(x,y) - \sum_{i=1}^n p[(x,y),(x_i,y_i)]\} \sum_{i=1}^n p[(x,y),(x_i,y_i)] \times (x - x_i)(i/n)^{k-1}$$

for the kth component in the gradient, $k = 1, 2, 3$. For $k = 4, 5, 6$ the last factor should be replaced by

$$(y - y_i)(i/n)^{k-4}$$

We code first the gradient, which takes most of the computing and we do it by brute force:

```
function grad=gradsnake1(id,a,n,c,temp,sigma,tau)
%computes gradient of posterior energy for snake patterns
%deformed digital inage id
%a=6-vector of similarity group
%n=number of generators
%c inverse of spreading deformation strength
%sigma=standard deviation in snakedef1
grad=zeros(1,6);
npoints=length(id(:,1));
v=[0:n-1]./n;
xs=temp(:,1)'+a(1)+a(2).*v+a(3).*v.^2;
ys=temp(:,2)'+a(4)+a(5).*v+a(6).*v.^2;
for x=1:npoints
        for y=1:npoints
x1=x/npoints;
y1=y/npoints;

                psv=exp(-c.*((x1-xs).^2+(y1-ys).^2));
                inner=id(x,y)-sum(psv);
                grad(1,1)=grad(1,1)+c*inner*sum(psv.*(x1-xs));
                grad(1,2)=grad(1,2)+c*inner*sum(psv.*(x1-xs).*v);
                grad(1,3)=grad(1,3)+c*inner*sum(psv.*(x1-xs).*v.^2);
                grad(1,4)=grad(1,4)+c*inner*sum(psv.*(y1-ys));
                grad(1,5)=grad(1,5)+c*inner*sum(psv.*(y1-ys).*v);
                grad(1,6)=grad(1,6)+c*inner*sum(psv.*(y1-ys).*v.^2);
                end
end
grad=-grad./(tau^2);
grad=grad+a./sigma^2;
```

This is not a well-designed program: it leads to excessive looping. It should be possible to speed it up by an order of magnitude.

The corresponding S.D.E. is completely straightforward:

```
function astar=restsnake1(id,n,c,temp,sigma,tau,dt,niter)
%carries out restoration of deformed image id
%dt=time step, niter=number of iterations
%result is a-vector
a=zeros(1,6);
rand('normal')
for t=1:niter
        wiener=rand(1,6);
        grad=gradsnake1(id,a,n,c,temp,sigma,tau)
        a=a-dt.*grad+(dt^.5).*wiener
end
astar=a;
```

Remark. The time step dt has to be made very small, say 10^{-7}, for the first few iterations, but can then be made larger. This was actually done in the following computer experiment in which we also left out the random terms. Even so it took an extremely long time to converge and one should be able to improve this a lot.

The deformed template in Figure 3 was widened (spread) into idsnake1 whose contour lines are shown in Figure 4.

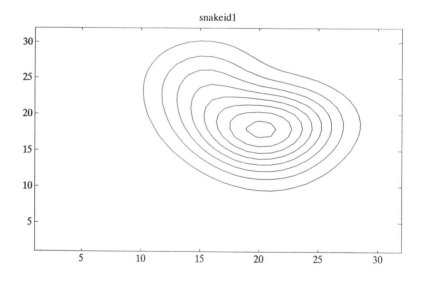

FIGURE 4. SNAKE IMAGE

Chapter 8: Computing Open Patterns

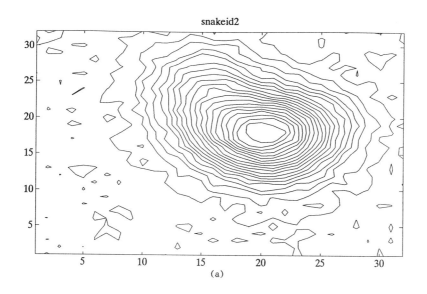

(a)

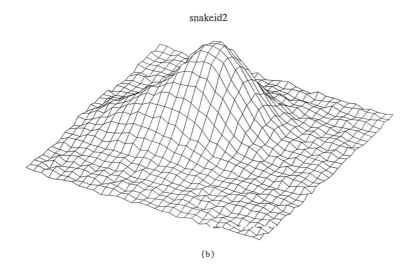

(b)

FIGURE 5. DEFORMED SNAKE IMAGE

174   *Chapter 8: Computing Open Patterns*

After that Gaussian noise was added we got $I^\mathcal{D}$ = snakeid2 with contours in Figure 5(a) and a mesh view in (b); the total deformation is seen to be severe. Applying restsnake1 we got the restored image in Figure 6. Compared with Figure 3, only a small part of the curve has been missed. Success!

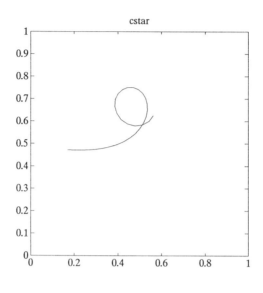

FIGURE 6

But the success is not as surprising as it may seem at first glance. The group deforming the template is very low dimensional, with only six parameters in contrast to other situations we are studying. This simplifies the task a great deal. Also, the computational effort was not reasonable.

The fact that we left out the random part in *restsnake1* means that we did MAP estimation (note 8.3) realized by a steepest descent algorithm of the most primitive type. This could of course be changed; the important thing to realize is, however, that by eliminating the random terms we could very well be stuck in a local energy minimum. In this case this does not seem to have happened, but it could, and one should be cautious when deciding whether to randomize the inference algorithm or not.

It is tempting to try the same method for the recognition of handwritten characters or numerals. Say, for simplicity, that we are dealing with a single numeral written with a pencil or pen. For each numeral $0, 1, \ldots 9$ choose one, or probably several, templates. For example, a 7 could be written with or without the European slash, as 7 or 7. Then we would introduce a deformation mechanism as a modification of the above; it should represent widening as well as noise due to imperfections in the paper-pen-pencil.

The resulting recognition algorithm would be impractical due to the massive computing effort needed, but this could no doubt be overcome. A more basic difficulty is the very choice of knowledge representation – does it really describe the way we write numerals? This is a question that requires subject matter knowledge, empirically based, that may be available but we have not exploited.

In general, to solve recognition tasks that humans do better than machines (at least today!) the crucial step is the creation of realistic knowledge representations. Enormous computing power cannot do it alone!

8.6. Computing weaves. In section 3.2 we looked at some common weave patterns and it seemed reasonable to classify them as closed since the value $I(x,y)$ depends upon its neighbors in a way that make the connector graph σ a lattice, hence with cycles. There is, however, another weave representation that is certainly closer to the physical mechanism, the loom plus warp plus filler, that produces the weave. Let us say the weave is going to be ℓx wide and ℓy long. We express the warp as an ℓx-vector taking values $1, 2, \ldots$ up to the number of colors used; it was 4 in this case. Similarly, we express the fillers by a vector of length ℓy, also taking as values the colors.

To experiment let us define warp and filler arbitrarily by the programs

```
function warp=setupwarp(wmat,lx)
%sets up symmetric warp for weaving of length lx=even number
%and transition probability matrix wmat
warp=markov1(1,wmat,lx/2);
warp=[warp,warp([lx/2:-1:1])];
```

and

```
function filler=setupfiller(fmat,ly)
%sets up symmetric filler for weaving of length ly=even   number
%transition probability matrix fmat
filler=markov1(1,fmat,ly/2);
filler=[filler,filler(ly/2:-1:1)];
```

For simplicity we have generated filler and warp at random using the utility markov1 from section 7.4. Note, however, that we force symmetry on these vectors by first generating the first half of the vectors and then completing them by concatenating the mirror image of the first half. We use the similarity group consisting of reflection right-left and up-down.

In these two programs the two last statements symmetrize the warp and filler around the midpoint of the weave. We shall do the same in the program setupharn below. The harnesses will be given as ℓy Boolean vectors of length ℓx. At the vertical (length) coordinate

y say that the harness is $h(y)$, for example

$$h(13) = [0\ 0\ 0\ 1\ 1\ 0\ 0\ 0\ 0\ 1]$$

if $\ell x = 10$. Again we generate these vectors a bit arbitrarily, organized into an $\ell y *$ ℓx Boolean matrix, by randomization. We start $h(1)$ as $[0\ 0\ 0\ ...\ 0]$ and then $h(2)$ for each component equal to the one in $h(1)$ with some probability 1-prob, otherwise the complementary value. This is done by the program setupharn, which also symmetrizes the harness matrix:

```
function harnesses=setupharn(lx,ly,prob)
%sets up symmetric harnesses for weaving ly long and lx wide
%both even numbers
%prob=change probability
harness=zeros(1,lx/2);
harnesses=harness;
rand('uniform')

for i=2:ly/2
        changes=rand(1,lx/2)<prob;
        harness=harness+changes;
        harness=modulo(harness,2);
        harnesses=[harnesses;harness];
end
harnesses=harnesses([[1:lx/2],[lx/2:-1:1]],:);
harnesses=harnesses(:,[[1:ly/2],[ly/2:-1:1]]);
```

The number of harnesses can be big, up to ℓy in number. This differs from real weaves where the number of harnesses is small and the reader is encouraged to modify setupharn and perhaps also setupwarp and setupfiller that are too random. Instead one could try deterministic functions for warp, filler, and harnesses; patterns of great beauty will sometimes be found but it takes a good deal of experimentation.

Now to the main function. The warp is raised or lowered at position (wide) x and at the yth filler, depending upon whether $h(x,y) = 0$ or 1. We can therefore write the color at x, y

$$I(x,y) = \begin{cases} w(x) \text{ if } h(x,y) = 1 \\ f(y) \text{ if } h(x,y) = 0 \end{cases}$$

when w is the warp vector, f is the filler and h is the harness matrix. This equation is implemented in line 12 of the function weave. The rest is really just displaying the colors.

In this way we generate the image $I(\cdot, \cdot)$ by the interaction of two images of connection type $\Sigma = $ LINEAR which are open, no cycles. This is related to the way Moiré fringes result from the interaction of the LINEAR structures, but a difference is that here the interaction is controlled by a third structure, the harness setup.

In Plate 9 we show a weave image obtained as described. Pretty! Beautiful patterns can be generated by this regular structure, especially by a reader who has access to MATLAB4 or some other software package with good color graphics. This offers exciting possibilities for the daring experimenter.

Chapter 9

Computing Closed Patterns

9.1. Computing Ising models. We have seen in section 6.1 one method, stochastic relaxation, for synthesizing Ising images. For their computing the main difficulty is Step 3–4. For this purpose we need the *env*-vector for a given site $= (x, y)$. Let us denote the current configuration by *cold*. Then *env* can be obtained as in line 7 below, which assumes periodic boundaries which accounts for the expressions $(y == nr) * nr$, and so on. Once env is known we can get the product of the four A-values (see section 6.1) as $\text{prod}(A(:, env + 1)')$; 1 is added since the rows of A are the numbers 1,2 rather than 0,1. Then we just apply the probsim utility we already have in section 7.4. The updating, at a single site spot, is then done (note 9.1) by

```
function cnew = updateising(spot,cold,A)
%author:Peter Schay,1992
x=spot(1)
y = spot(2);
[nr,nc] = size(cold);
cnew = cold;
env = [cnew(x,y+1-((y==nr)*nr)), cnew(x-1+((x==1)*nc),y),
cnew(x,y-1+((y==1)*nr)),cnew(x+1-((x==nc)*nc),y)];
prob = prod(A(:,env+1)');
prob = prob/sum(prob);
cnew(x,y) = probsim(prob)-1;
```

To implement the other steps of the stochastic relaxation is actually easier, although the following function will be longer since some switches are included to allow the user several options. The program first interrogates the user about parameters and choices as described, and initializes $c(0)$ as the user has requested. It also displays $c(0)$. The main body of the program selects a site, again in a way asked for. Then the function updateising is called in the line cnew = updatising (...) and iterations are continued until pauseiters is answered by 0.

Chapter 9: Computing Closed Patterns

```
function cnew = isingmodel()    % by Peter Schay
  cnew = [];
% get some parameters of the model from the user:
  vec = input('Enter [rows, cols] of the pattern matrix -->   ');
  nr = vec(1);
  nc = vec(2);
  pauseiters = input('Display pattern after how many updates -->');
  vec = input('Enter [b,c]   A(0,1)=b and A(1,1)=c -->');
  b = vec(1);
  c = vec(2);
  inpat = input('Initial pattern (0=zeros 1=ones 2=uniformly random ) -->');
  smeth = input('Sweep method (0=inorder 1=randomorder) -->');
  numhatches = input('Hatches per display square -->');
% set up the initial pattern:
  for i=1:nr,
    if inpat == 0,
       cnew=[cnew;zeros(1,nc)];
    elseif inpat == 1,
       cnew=[cnew;ones(1,nc)];
    elseif inpat == 2,
       rand('uniform');
       cnew=[cnew;rand(1,nc)>.5];
    end;
  end;
  displaypat(cnew,numhatches);
  junk = input('Hit return to continue');
  A = [1 b; b c];
% locally update the pattern and display it after <pauseiters> updates:
  iters = 0; quitting = 0;
% the following loop generates all of the (row,col) pairs in the pattern
% matrix, except for those on the boundaries. Example:a   4x5 (nr x nc)
% has rows = [2 2 2 3 3 3]
% and cols = [2 3 4 2 3 4]. Matching elements, we have (2,2), (2,3) etc.
% indexlist= [1 2 3 4 5 6] is a systematic "inorder" traversal of the
% points to be updated. indexlist can be also be in randomorder, which
% is done by just getting the list of indices returned from a sort of
% random numbers. (example: indexlist = [3 5 2 1 4 6])
  rows = []; cols = [];
  for i = 1:nr,
    rows = [rows,[i.*ones(1,nc)]];
    cols = [cols,[1:nc]];
  end;
  numelems = nr*nc;
  indexlist = [1:numelems]; % default index ordering is inorder
% here comes the outer loop to update every pattern element pauseiter times
% and then display the new pattern:
  while quitting == 0,
    if smeth == 1, % reshuffle the index ordering if we're in randomorder:
       [junk,indexlist] = sort(rand (1,numelems));
    end;
% update every element in the pattern matrix:
    for i = 1:numelems,
       cnew = updateising([rows(indexlist(i))
cols(indexlist(i))], cnew, A);
    end;
% display cnew if we've finished the required (pauseiters) iterations:
    iters = iters + 1;
    if iters >= pauseiters,
       displaypat(cnew,numhatches);
       sprintf('Performed %g iterations',iters)
       iters = 0;
       pauseiters = input('further iterations to perform (0 to quit) -->');
       if pauseiters == 0, quitting = 1;
       end;
    end;
  end
```

We can now go ahead to image restoration. To implement $\mathcal{D}$ is almost immediate; we just flip the 0,1 values with probability ϵ:

```
function cdistorted = distort(cold, epsilon)
  cdistorted = cold;
  [nr,nc] = size(cold);
  for i = 1:nr,
    for j = 1:nc,
      % flip cold(i,j) with prob epsilon:
      cdistorted(i,j) = abs(cold(i,j)-(rand(1)<epsilon));
    end;
  end;
```

Here cold means the c (or I) obtained from the Ising model and c distorted is our $c^{\mathcal{D}}$ (or $I^{\mathcal{D}}$).

The restoration code is very much like Ising model synthesis except that the switches have been left out, making the program shorter.

```
function cnew = restoreising(cdistorted,A,ep,niter)
  cnew = cdistorted;
  [nr,nc] = size(cnew);
  numelems = nr*nc;
  for it = 1:niter,
    rows = []; cols = [];
    for x = 1:nr,
      for y = 1:nc,
        env = [cnew(x,y+1-((y==nr)*nr)),
cnew(x-1+((x==1)*nc),y),
cnew(x,y-1+((y==1)*nr)),cnew(x+1-((x==nc)*nc),y)];
        prob =
prod([A(:,env+1)';abs(cnew(x,y)-ep),abs(1-ep-cnew(x,y))]);
        prob = prob/sum(prob);
        cnew(x,y) = probsim(prob)-1;
      end;
    end;
  end;
```

Here we sweep the picture like a TV scan, see lines 7,8 and the environment vector env is obtained as before. In lines 9,10 we also include the factor ϵ or $1 - \epsilon$ as the case may be. For cnew we have to subtract 1 from probsim(prob) to get generator values = 0 or 1.

This code was used for synthesis and restoration in section 6.1. As emphasized above, once pattern synthesis has been well organized analysis required little modification in principle (but much more computing may be needed).

9.2. Computing boundary models. To experiment with boundary models of the type introduced in section 6.2 does not require any drastic changes in the programs. To illustrate this let us use generators given in Figure 6 of section 6.2. The main difference compared to the Ising synthesis/analysis is that factors $A(g_1, g_2)$ will now be replaced by

Chapter 9: Computing Closed Patterns

factors $A[\beta_1(g_1), \beta_3(g_2)]$ and so on. Otherwise the code is almost the same. Remember that the first entries in the Q-vector should be made much larger than the others.

In Figure 1a we show a synthesized configuration where B stands for a boundary generator (with some orientation from BSG), * means an inside point and the blanks are outside points. In Figure 1b the corresponding image is shown.

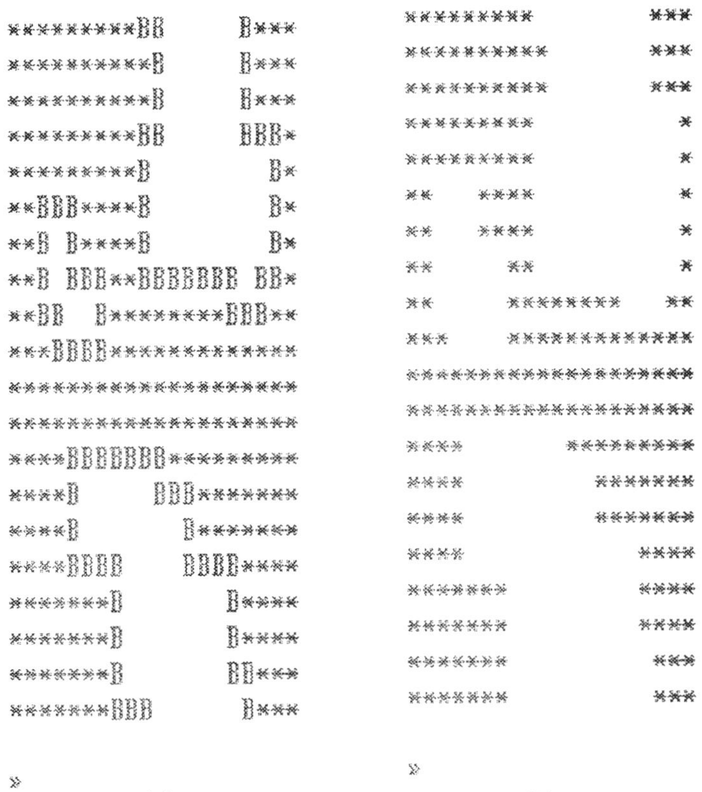

FIGURE 1. BOUNDARY PATTERNS

We then deform (b) by 10% noise and get Figure 2a. Applying a restoration algorithm very similar to that of the previous section we get I^* shown in Figure 2b. It should be compared to Figure 1b: the result is excellent.

The generators used for boundary patterns are more (but not much more) structured than those for the Ising model. For successful applications of pattern theory it is necessary

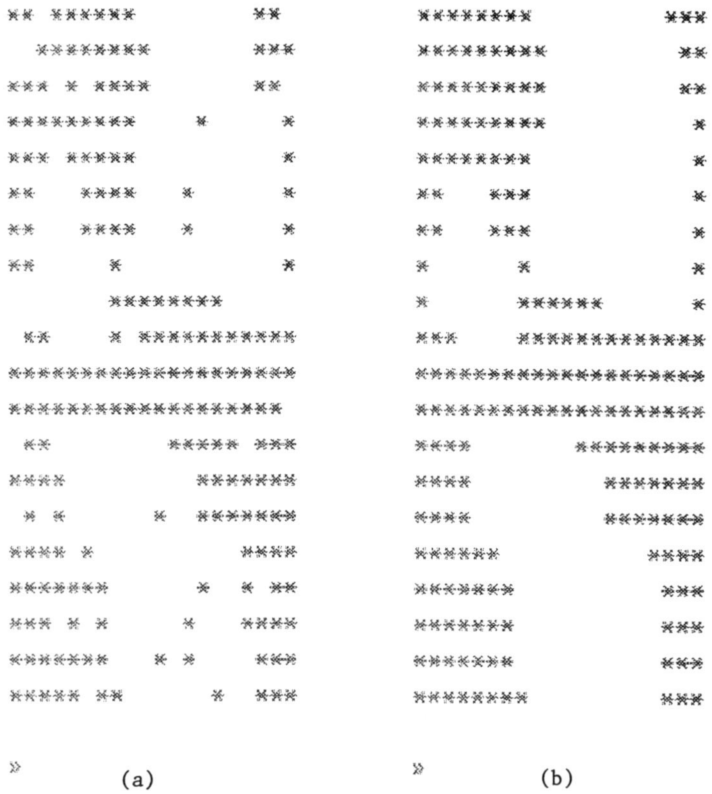

FIGURE 2. BOUNDARY PATTERNS

to include enough subject matter knowledge in the representations by structuring G and Σ in a realistic way.

9.3. Computing deformable templates. To implement deformable templates as in section 6.3, curves in $\mathbb{R}^2$, the main task is to simulate the v-sequence by

$$v = F \begin{pmatrix} e_1 \\ e_2 \\ \vdots \\ e_n \end{pmatrix}$$

where the e_i's form an i.i.d. sample from $N(0, \sigma^2)$. The latter is straightforward in MATLAB with rand$(1, n)$ under the option rand("normal").

Now F is the inverse of the square root N of M, $N^2 = M$, where M is the matrix of the quadratic form

$$Q = v^T M v = \sum_{i=1}^{n} (v_{i+1} - av_i)^2; \; v_{n+1} = 1.$$

It was seen that M was a Toeplitz matrix so that

$$M = \text{toeplitz}([2, -a, zeros(1, n-3, -a)])$$

and

$$F = inv(M \wedge .5).$$

Lines 7–14 do this for v as well as u; in the latter case we add 1 to all its components to make them have mean value 1. In lines 15–21 we compute the generators g_i^0 of the template whose vertices are represented by coordinates in basis, etc. Then we can get the g_i's, in the code gs as a two column matrix, by multiplying the g_i^0 vector by Δ_i expressed in the u_i, v_i. After this we need only form the cumulative sums

```
function config=synthdt(ksis,etas,a,sigma)
%synthesizes deformed template with vertices ksis,etas
%by direct (non-iterative) method
%a=coupling coefficient;sigma=variability coefficient
%applies to closed curves in2D
n=length(ksis);
rand('normal');
M=toeplitz([2,-a,zeros(1,n-3),-a]);
N=M^.5;
F=inv(N);
us=F*rand(n,1);
us=1+sigma.*us;
vs=F*rand(n,1);
vs=sigma.*vs;
ksi=[ksis,ksis(1)];
eta=[etas,etas(1)];
g0xs=ksi([2:n+1])-ksi([1:n]);
g0ys=eta([2:n+1])-eta([1:n]);
g0s=zeros(n,2);
g0s(:,1)=g0xs';
g0s(:,2)=g0ys';
gs=zeros(n,2);
for i=1:n
        gs(i,:)=([us(i),vs(i);-vs(i),us(i)]*(g0s(i,:))')';
end
sumgs=sum(gs);
gs(:,1)=gs(:,1)-(1/n)*sumgs(1,1);
gs(:,2)=gs(:,2)-(1/n)*sumgs(1,2);
config=cumsum(gs);
```

$$config = cumsum(gs)$$

after having modified the gs to satisfy the closure condition. If the template is not a closed curve the latter is not needed.

When experimenting with the program synthdt it is tricky to choose the parameters a (for stochastic coupling) and σ (for variability) in order to produce reasonable looking deformations. To help with this one could calculate, for various values of a and σ the covariance matrix of the us (as well as the vs)

$$R = \sigma^2 M^{-1}$$

and plot, for example, $R(1,1)$ as a function of a, given σ, and also the correlation sequence $R(1,i)./R(1,1)$ for given a. This will tell the experimenter how the stochastic similarity group elements are expected to behave. Figure 17 in section 6.3 shows an example of a synthetic picture.

When we go to background deformations in 2-D we can use the series expansions in section 6.3 noting that the eigen functions of Δ can be written in terms of sines and cosines. Now is the time to discretize and we do it on an $m \times m$-lattice; see line 7 in the program deform 2. In its arguments power stands for the exponent p in section 6.3, sigma is the standard deviation of the e-fields and order is the largest k and ℓ value used in the expansion. The choice of order is of interest in itself; it is not just to limit computing that we make it considerably smaller than m (note 8.5).

We first simulate the u, v-fields by successively adding $\varphi_{k\ell}$-values in the k and ℓ loop starting at line 14. The choice of sine and cosine is due to the mixed boundary conditions we use but this is of little importance. The two last statements multiply u and v by a constant σ in order to be able to vary the noise power of the e-fields

```
function [u,v]=deform2(m,order,sigma,power)
%computes 2D-deformation using eigen-vectors
%of Laplacian with mixed boundary conditions
%on m*m lattice and order*order eigen vectors used
%sigma is noise power
rand('normal')
angles=(pi/m).*[0:m];
u=zeros(m+1,m+1);
v=u;
lambdas=outer(([0:order].^2),'+',([1:order].^2));
lambdas=lambdas.^(-power/2);
zs1=rand(order+1,order);
zs2=rand(order+1,order);
for k=0:order-1
        for l=[1:order]
                sins=sin(k.*angles);
                coss=cos(l.*angles);
                u=u+zs1(k+1,l).*lambdas(k+1,l).*(sins'*coss);
                v=v+zs2(k+1,l).*lambdas(k+1,l).*(coss'*sins);
        end
end
u=u.*sigma;
v=v.*sigma;
```

Once we have all the u's and v's it is an easy matter to find the deformed x, y values: $x+u(x,y)$ and $y+v(x,y)$. This is done in the straightforward program comp2d in which we have discretized the unit square into an $m \times m$ lattice so that x's and y's are all multiples of $1/m$ before the deformation.

```
function [xs,ys]=comp2d(m,u,v)
%computes deformation fields xs and ys from
%displacement fields u and v
%on m*m-lattice
xs=u+(1/m).*[0:m]'*ones(1,m+1);
ys=v+(1/m).*ones(1,m+1)'*[0:m];
```

We display the deformed lattice by drawing line segments between points given by the xs and ys array; see lines 9 and 14, both vertically and horizontally.

```
function seedef2(m,xs,ys)
%plots deformation fields xs,ys obtained from
%functions deform2 and comp2d
%on m*m-lattice
clg
hold on
for k=1:m
        for l=1:m+1
                plot([xs(k,l) xs(k+1,l)],[ys(k,l) ys(k+1,l)]);
        end
end
for k=1:m+1
        for l=1:m
                plot([xs(k,l) xs(k,l+1)],[ys(k,l) ys(k,l+1)]);
        end
end
```

The reader should experiment with deform2, especially by choosing different values for *order*. The effect is quite noticeable and is related to how "local" we want the deformations to be. Look also at Figure 9 in section 6.3.

The above was for 2-D but the physical world we live in is 3-D. In principle there is no difficulty in extending this pattern theoretic model to 3-D but actual applications require extensive data collection/analysis. A daring attempt is under way to do this for medical imaging creating a "digital anatomy" where normal as well as pathological anatomies are represented by deformable templates in the same spirit as above. If this succeeds it will lead to powerful tools to help diagnosticians in various medical specialities.

9.4. Star-shaped patterns. Let us now apply the ideas of deformable templates to star-shaped patterns. The generators will now be chosen as radii $r_i \leq ; i = 1, 2, \ldots n$, from the origin and with equally spaced angles $\varphi_i = \frac{2\pi i}{n}, i = 1, 2, \ldots n$.

Let us compute a lobed template, say with the number of lobes $= m$ and with

$$r_i = \alpha + \beta \cos \frac{2\pi i m}{k}.$$

Here α and β should be positive parameters with $0 < \alpha - \beta < \alpha + \beta < 1$. The r's are computed by the self-explanatory

```
function rs=star(nangles,nlobes,alpha,beta)
%computes vector rs of radii in polar coordinates
%of star shape with nangle angles and nlobes lobes
%and positive parameters with alpha+beta<1 and alpha-beta>0
fis=(2*pi/nangles).*[0:nangles-1];
rs=alpha+beta.*cos(nlobes.*fis);
```

where nangles plays the role of n and nlobes that of m. A star-shaped template is shown in Figure 3 obtained by seestar(rs):

```
function seestar(config)
%displays star shape given by radii in config
n=length(config);
angles=(2*pi/n).*[0:n-1];
xs=config.*cos(angles);
ys=config.*sin(angles);
xs=[xs,xs(1)];
ys=[ys,ys(1)];
clg
axis([-1 1 -1 1]);
plot(xs,ys)
```

Designing the prior density over $\mathcal{C}(\mathcal{R})$ is more challenging. Since the generator space can be imbedded in the interval $[0, 1]$ (leaving out for the moment the angular information) we should let the elements s of the similarity group S map this interval onto itself. A simple choice is

$$s : x \mapsto x^s$$

where s is a positive real number, so that the composition $S_1 S_2$ means $s_1 + s_2$ with obvious identification of the group element with the real number that represents. We shall let the angle φ in $g = (r, \varphi)$ be left unchanged by the similarities.

On S, or equivalently $\mathbb{R}_+$, we introduce a Γ-probability density $p(s) \propto s^p e^{\gamma s}, s > 0, p$ and γ positive parameters. To find the maximum of $p(\cdot)$ we solve

$$0 = \frac{d \log p(s)}{ds} = \frac{d}{ds}(\text{constant} + p \log s - \gamma s) = \frac{p}{s} - \gamma$$

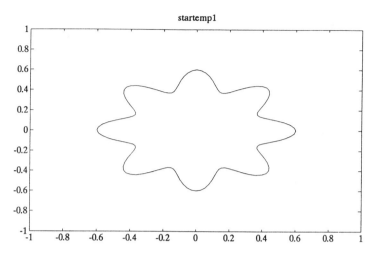

FIGURE 3. STAR-SHAPED TEMPLATE

so that we choose $p = \gamma$ in order to have the mode of $p(\cdot)$ located at $s = 1$, which is the identity element of our group.

We also need a coupling term in the energy function and we shall choose it the same way as in the previous section, so that the total energy function of the configuration $c = \text{CYCLIC}(s_1 g_1^0, g_2 g_2^0, \ldots s_n g_n^0)$ (where the g_i^0 are the generators of the template) becomes

$$E(c) = \sum_{i=1}^{n}\{\gamma s_i - p \, log \, s_i + \frac{b}{2}\sum_{i=1}^{n}(s_{i+1} - s_i)^2\}.$$

For pattern synthesis we shall use the general S.D.E. from section 6.6 and we then need the gradient of the energy with components

$$\frac{\partial E(c)}{\partial s_i} = \gamma - \frac{p}{s_i} + b(-s_{i-1} + 2s_i - s_{i+1}).$$

We can then code the S.D.E. into a simple program:

```
function config=synthstar(rs,gamma,b,dt,sigma,niter)
%synthesizes star shape from template configuration rs
%with gamma distribution, parameter gamma
%coupling coefficient b ,Wienerprocess constant sigma
%solves C.D.E. with niter iterations
%time step=dt
n=length(rs);
ss=ones(1,n);
rand('normal')
for t=1:niter
        shiftssl=ss([[2:n],1]);
        shiftssr=ss([n,[1:n-1]]);
        wiener=rand(1,n).*sigma*dt^.5;
        grad=gamma.*ones(1,n)-gamma./ss+b.*(-shiftssl+2.*ss-shiftssr);
        ss=ss-dt.*grad+wiener;
end
config=rs.^ss;
```

Note that in the last line we go from the vector ss, meaning $(s_1, s_2, \ldots s_n)$ to the radii vector rs of the generators. This is as before because the relation $g_i^0 \to s_i g_i^0 = g_1$ is bijective. An example is shown in Figure 4 which should be compared to the template in Figure 3.

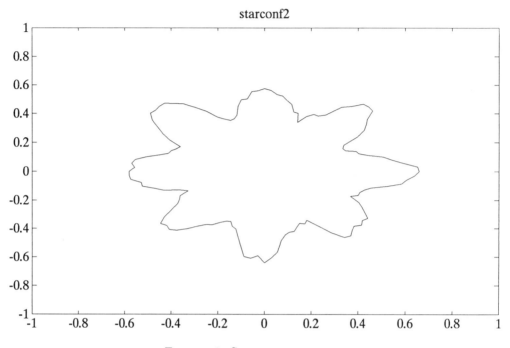

FIGURE 4. STAR-SHAPED IMAGE

Now we shall introduce a fairly natural deformation mechanism $\mathcal{D}$ of the type "heterogeneous Poisson process." By this we mean that on each of the equally spaced radii of length 1 beginning at the origin of the polar coordinate system, we simulate a Poisson process that has intensity $r\lambda_{im}$ for $0 < r < r_i$ and intensity $r\lambda_{out}$ for $r_i < r < 1$; this is for the ith generator. The motivation for including the factor r in the intensities is that the area element in polar coordinates is $r dr d\varphi$ and we have a spatial, not linear, Poisson process in mind, although we have discretized angles for simplicity.

In the following program defstar, that implements $\mathcal{D}$, we discretize each radius of length 1 into npoints points, perhaps with npoints = 100 or so. We then have to rescale the Poisson intensities by dividing them by npoints. We compute n_{in} = the number of these npoints points inside r_i, and define a vector intensity whose k^{th} entry is λ_{in} if $r_i < k/\text{npoints}$ and λ_{out} else. Then we have to simulate a Bernoulli process (see note 9.3), which is done applying our utility select (see section 7.4.5) to the vector boole which has 1's and 0's indicating whether the Bernoulli event occurred or not. We do this for each generator (ray) and concatenate new points, if there are any, successively to the 2-column matrix id that means the deformed image $I^\mathcal{D}$. This is done by the program defstar:

```
function id=defstar(nangles,rs,lambdain,lambdaout,npoints)
%heterogeneous Poisson deformation mechanism for star shapes
%radii of shape=rs
%number of angles=nangles
%Poisson intensities lambdain and lambdaout
%deformed image=id
%radius of length 1 discretized into npoints points
radii=[1:npoints]./npoints;
lambdain=lambdain/npoints;
lambdaout=lambdaout/npoints;
rand('uniform');
for i=1:nangles
        nin=floor(rs(1,i)*npoints);
        intensityin=lambdain.*ones(1,nin);
        intensityout=lambdaout.*ones(1,npoints-nin);
        intensity=[intensityin,intensityout];
        intensity=intensity.*radii;
        random=rand(1,npoints);
        boole=random<intensity;
        choosers=select(radii,boole);
        number=length(choosers);
        if number>0,
        xs=choosers.*cos(2*pi*i/nangles);
        ys=choosers.*sin(2*pi*i/nangles);
        mat=zeros(number,2);
        mat(:,1)=xs';
        mat(:,2)=ys';
        id=[id;mat];
    else
    end
end
```

A deformed image, shown in Figure 5, is obtained by executing seeid1(id), a utility from section 7.4.3. The lobes are now hidden by noise, but can still be seen fairly well.

To organize image restoration along the principles of Chapter 6 note that the likelihood factor L can now be written as a product over $i = 1, 2, \ldots n$ where the ith factor is proportional to

$$\lambda_{in}^{n_{in}(sr)} \lambda_{out}^{n_{out}(sr)} \, exp - [\lambda_{in} sr + \lambda_{out}(1-sr)]$$

where
$$\begin{cases} n_{in}(x) = \text{number of points in } I^\mathcal{D} \text{ along ray } i \text{ up to distance } x \\ n_{out}(x) = \text{number of points in } I^\mathcal{D} \text{ along ray } i \text{ between } x \text{ and } 1 \end{cases}$$

and r is the radius associated with g_i^0 and s the corresponding similarity.

Hence, with obvious notation

$$log \, L = \sum_{i=1}^{n} [n_{in}^i(s_i r_i) log \, \lambda_{in} + n_{out}^i(s_i r_i) log \lambda_{out} - \lambda_{in} s_i r_i - \lambda_{out}(1 - s_i r_i)].$$

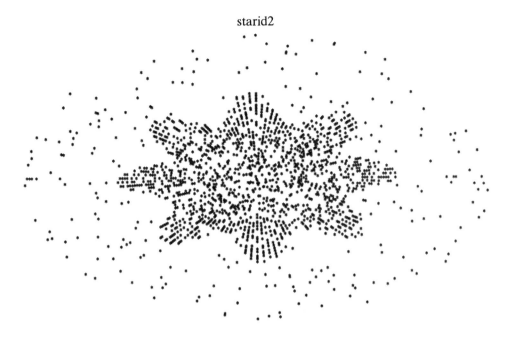

FIGURE 5

The gradient with respect to the vector $s = (s_1, s_2, \ldots s_n)$ will then have the ith component

$$log \lambda_{in} \frac{dn^i_{in}(s_i r_i)}{ds_i} + log \lambda_{out} \frac{dn^i_{out}(s_i r_i)}{ds_i} - r_i(\lambda_{in} - \lambda_{out})$$

but this has to be interpreted carefully, since the n^i's are integer valued and hence their derivatives not immediately meaningful. Instead we shall interpret the first of the two derivatives as divided differences. We can then set up an S.D.E. for restoration, but will not do so here; the reader will find it challenging!

9.5. Multimodal inference. Let us keep the setup in section 6.3 but change it in one respect that will lead to a radically different view of the logic that underlies pattern inference. This is a difficult but intellectually rewarding topic!

As before, let the configuration be $c = \text{CYCLIC}(g_1, g_2, \ldots g_n)$ where the vectors $g_i = (g_{ix}, g_{iy})$ are associated with the vertices $z_i = (x_i, y_i)$ by $g_i = z_{i+1} - z_i$. Now, however, we shall use two (or more) templates

$$\begin{cases} c^1 = \text{CYCLIC}(g^1_1, g^1_2, \ldots g^1_n) \\ c^2 = \text{CYCLIC}(g^2_1, g^2_2, \ldots g^2_n) \end{cases}$$

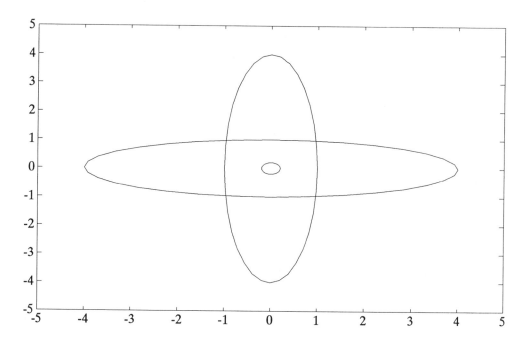

FIGURE 6. INITIAL AND TEMPLATE CONFIGURATIONS

and use an acceptor function $A(g',g'') = exp[-E(g',g'')]$ in exponential form (see section 4.1) where the energy will be given the form

$$E(g',g'') = \frac{1}{2}E_{auto}(g') + \frac{1}{2}E_{auto}(g'') + E_{couple}(g',g'')$$

where the coupling energy will be of the same form as before

$$E_{couple}(g',g'') = b\|g'-g''\|^2 = b[(g'_x - g''_x)^2 + (g'_y - g''_y)^2].$$

The auto energy, however, will be quite different

$$\begin{aligned}E_{auto}(g) &= a[\|g-g'\|^2\|g-g''\|^2]\\&= a[(g_x - g^1_x)^2 + (g_y - g^1_y)^2][(g_x - g^2_x)^2 + (g_y - g^2_y)^2]\end{aligned}$$

which is to be understood as follows. For $g = g_i$ we mean $g^1 = g^1_i$ and $g^2 = g^2_i$ from the two templates. Note the multiplicative form in E_{auto} which implies that minimum energy (maximum probability density) is attained both for $g = g^1$ and for $g = g^2$; it is bimodal (or multimodal if several templates are introduced).

To get an idea of how the resulting random configurations will behave let us choose g^1 and g^2 as ellipses with half axes (1,4) and (4,1) respectively, and let us initialize the configuration by $c = c^0$ as a small circle. They will look like Figure 6.

In the stochastic differential equation, motivated as before, we need the gradient of the total energy. Since the total configuration energy can be written as (with $i = n + 1$ understood as 1)

$$E(c) = a \sum_{i=1}^{n} [(g_{ix} - g_{ix}^1)^2 + (g_{iy} - g_{iy}^1)^2][(g_{ix} - g_{ix}^2)^2 + (g_{iy} - g_{iy}^2)^2]$$

$$+ b \sum_{i=1}^{n} [(g_{i+1x} - g_{ix})^2 + (g_{i+1y} - g_{iy})^2]$$

the gradient, that will be represented by a $2 \times n$ matrix called "grad" in the code, will have the x-components given through

$$\frac{1}{2} \frac{\partial E(c)}{\partial g_{ix}} = a(g_{ix} - g_{ix}^1)Q_2 + a(g_{ix} - g_{ix}^2)Q_1$$

$$+ b[-g_{i-1x} + 2g_{ix} - g_{i+1x}]$$

with the two quadratic forms

$$\begin{cases} Q_1 = \sum_{i=1}^{n} \|g_i - g_i^1\|^2 \\ Q_2 = \sum_{i=1}^{n} \|g_i - g_i^2\|^2. \end{cases}$$

The expression for the y-components of "grad" is quite similar.

To compute "grad" we write the function "gradpot":

```
function           grad=gradpot(g,g1,g2,a,b)
%computes gradient  of bimodal potential
%with coupling terms;a and b are respective
%proportionality constants
%all g's are 2-row matrices
length=size(g);
n=length(2);
sq1=sum((g-g1).^2);
sq2=sum((g-g2).^2);
shiftgr=g(:,[[2:n],1]);
shiftgl=g(:,[n,[1:n-1]]);
grad=zeros(2,n);
grad(1,:)=a.*((g(1,:)-g1(1,:)).*sq2+(g(1,:)-g2(1,:)).*sq1);
grad(1,:)=grad(1,:)+b.*(-shiftgr(1,:)+2.*g(1,:)-shiftgl(1,:));
grad(2,:)=a.*((g(2,:)-g1(2,:)).*sq2+(g(2,:)-g2(2,:)).*sq1);
grad(2,:)=grad(2,:)+b.*(-shiftgr(2,:)+2.*g(2,:)-shiftgl(2,:));
```

Here $sq1, sq2$ correspond to the quadratic forms Q_1, Q_2 and shiftgr, shiftgℓ give the generators g_{i+1} and g_{i-1} respectively, shifted right and left respectively.

The stochastic differential equation then is easily computed by

```
function gnew=growl(ginit,g1,g2,a,b,dt,niter,sigma)
%growth with bimodal exponent;g's are sides (generators)
%gnew are resulting sides; apply vert to get vertices
n=length(ginit(1,:));
g=ginit;
rand('normal');
for t=1:niter
        grad=gradpot(g,g1,g2,a,b);
        gnew=g-dt.*grad+(sigma*dt^.5).*rand(2,n);
        gnew(1,:)=gnew(1,:)-ones(1,n).*sum(gnew(1,:))/n;
        gnew(2,:)=gnew(2,:)-ones(1,n).*sum(gnew(2,:))/n;
g=gnew;
end
```

where g_{init} is the initial configuration (some abuse of notation) and $g1$ and $g2$ mean the templates c^1 and c^2. The time step is called dt, niter is the number of iterations and sigma is the standard deviation in the Wiener process term.

Running this program and plotting we get thought-provoking pictures like the one in Figure 7; the little circle is c_{init}. It is clear what has been happening. For small values of σ (low temperature) the generators will be attracted either to the g_i^1's or g_i^2's, whichever is closest. But this depends on how we start the synthesis, on the choice of cinit. With the circular shape the choice will change drastically as i goes from 1 to n, which explains the four main lobes in Figure 7.

When the algorithm makes this choice it does not compromise between the two alternatives by some averaging procedure; instead it decides on one or the other (or close to them). The reason is the bimodal form of the energy function. If it had been convex, as was the case in sections 9.3 and 9.4, a unique minimum exists and a minimum problem then has an averaging solution. This is related to harmonic functions for which a mean value theorem always holds.

This pattern synthesis is a highly specialized example of a principle of great generality and scope. Forget for a moment the particular setup and abstract to a less restricted situation. If the energy function is multimodal (and hence non-convex) the synthesis and, more importantly, the pattern inferences can be of uncompromising type; the logic will not try to satisfy all the requirements and data elements by averaging. Instead it will satisfy some of the requirements very well and neglect others.

We could sum up this as follows. Such logical inferences have three characteristics:

a) they are carried out in *parallel* (not in series) as described by the connector graph;

b) they are *uncertain* rather than deterministic due to the stochastic elements;

and, most importantly

c) they are not compromising (averaging) but *alternative*.

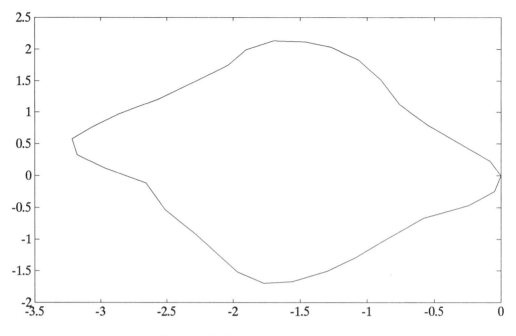

Figure 7. Bistable configuration

This type of pattern inference deserves a deeper study and extension to more general situations. It has some characteristics of real human reasoning that traditional logic lacks.

9.6. Computing character string restoration. To implement the method of section 6.4 is harder than some of the earlier tasks but synthesis of I and definition into $I^{\mathcal{D}}$ is fairly straightforward. We do it in two steps. In the first one we simulate the string $m_1, m_2, \ldots m_i, , \ldots m_n$, one at a time so that

$$P(m_i = k) = p_k$$

where we call (p_0, p_1, p_2) by the name ps. Once that is done (in the first for-loop below) we just calculate I by applying m_i to $x_i = i^{th}$ entry of template and concatenate the results. When I has been computed we corrupt it by noise on the error level ϵ; this is done in the second for-loop.

Remark. As the reader will notice, the error mechanism is not exactly as in section 6.4 since the actual error probability in the program is not ϵ but

$$\epsilon(1 - 1/nrspace).$$

This can be changed easily.

This is implemented by the program synthstr1

```
function [id,image]=synthstr1(template,nxspace,ps,epsilon)
%synthesizes numerical strings from X={1,2,...nxspace}
%from string template by applying m-transformations to
%individual entries in template with probabilities
%given by 3-vector ps
%then deforms resulting image by noisy
% channel with error probability epsilon
%errors uniformly distributed over X
n=length(template);
image=[];
for i=1:n
        string=template(i);
        number=probsim(ps);
        newstring=string(1,ones(1,number-1));
        image=[image,newstring];
end
limage=length(image);
id=image;
for i=1:limage
        error=probsim([epsilon,1-epsilon]);
        if(error==1)
        number=probsim((1/nxspace).*ones(1,nxspace));
        id(i)=number;
        else
        end
end
```

Coding the restoration algorithm takes more thought. Look at the code that follows; line 11 initializes the m-string. We separate three cases. If $\ell(=\ell id) = n$ (see section 6.4 for notation) we make the natural choice: all $m_i = \ell$. If $\ell < n$ we choose m as ℓ 1's followed by $n - \ell$ zeros. If $\ell > n$ we choose the first t m_i-values as 1 and the remaining s as 2. We have to satisfy

$$\begin{cases} n = t + s \\ \ell = t + 2s \end{cases}$$

so that $s = \ell - n$ and $t = n - s$. This is arbitrary and it may be possible to find a better initialization.

Now we start the iteration loop. Once the site i has been selected, line 23, we compute the

```
function istar=reststr1(template,nxspace,ps,epsilon,id,niter)
%restores deformed string image id with the result istar
%see comments in function synthstr1 for meaning of parameters
%niter is number of iterations
n=length(template);
lid=length(id);
unif=(1/(n-1)).*ones(1,n-1);

start=[0 0 1 2];
stop=[1 2 2 2];
lprobs=[2 3 2  1];
%first initialize the ms-vector for three cases
        if(lid==n)
                ms=ones(1,n);
        elseif(lid<n);
                ms=[ones(1,lid),zeros(1,n-lid)];
        else
                s=lid-n;
                t=n-s;
                ms=[ones(1,t),2.*ones(1,s)];
        end
%start iterations
for t=1:niter
        i=probsim(unif);
        lsub=ms(i)+ms(i+1);
        if(lsub>0);
                before=sum(ms(1:i-1));
                ystring=id(before+1:before+lsub);
                probs=zeros(1,4);
                        for k=start(lsub):stop(lsub);
                        x1=template(i);
                        x2=template(i+1);
                        xstring=[x1(1,ones(1,k)),x2(1,ones(1,lsub-k))];
                        probs(k+1)=ps(k+1)*ps(lsub-k+1);
                        nfit=sum(xstring==ystring);
                        probs(k+1)=probs(k+1)*((1-epsilon)^nfit)*epsilon^(lsub-\
nfit);

                end
                probs=probs./sum(probs);
                number=probsim(probs);
                ms(i)=number-1;
                ms(i+1)=lsub-ms(i);
                else
                end
end
istar=[];
for i=1:n
        g=template(i);
        istar=[istar,g(1,ones(1,ms(i)))];
end
```

length ℓ_i (ℓ sub) $= m_i + m_{i+1}$. Let us think for a moment about what are possible cases. Ruling out the case ℓsub $= 0$, when no change can be made, we have ℓsub $= 1, 2, 3, 4$. What are the possible combinations m'_i, m'_{i+1}? We get the table

ℓsub =	1	2	3	4
	0,1	0,2	1,2	2,2
	1,0	1,1	2,1	
		2,0		

This is coded into lines 8, 9, 10 of the program. Here the start-vector means the first possible m'_i-value, stop means the last possible m'_i-value, and ℓprobs means the number of possible values for given ℓsub.

Now we can find the y-string that will be needed to compute the conditional probabilities. The j-values of y_j are computed in lines 26, 27. We shall compute a 4-vector prob of conditional probabilities, although only ℓprobs(ℓsub) of them will be positive. This is done by first finding the x-string to be deformed and then filling in the positive values in probs looping from start(ℓsub) to stop(ℓsub). In line 35 we multiply by the factors $\epsilon^{N_e}(1-\epsilon)^{\ell-N_e}$ when the number N_e of errors was found in the previous line. After normalizing probs we simulate it using the utility probsin from Chapter 7, and get the new values for m_i, m_{i+1}. Once the m-value is obtained we find the restored image iter by direct computation.

Let us try this on the template

$$aaabbccccdddeeeeeeefffggghhhihhgfebbabcddeeeffffgggggghhh$$

with ps = [.2 .6 .2] and get the image I, a little longer than the template,

$$aaaaaabbbbcccdddeeeeeeffffggghhhihhggfebbbabddeeeffffgggghhh.$$

Now we deform I by severe noise, $\epsilon = 50\%$ and get the chaotic string

$$dfghbagbbhcfediceeeeeefffffigdabhfhhgbfabdgbbidfeddifcaeggggaih.$$

The restored string iter is, after niter = 1000

$$aabbcccdddddeeeeeefffggggghhhhiihhgfeebabceeeeffffffgggggggghhhhh.$$

To visualize this we show I and I^D (dotted line) in Figure 8. In Figure 9 we show I and I^* (dotted line) and it is seen that much of the noise is removed. I^* is however shifted a bit to the left. This is probably due to the fact that the information in the template is given much weight due to the high value of ϵ. If we had run reststr1 with a smaller ϵ the shift would perhaps be less pronounced but more of the noise would remain. One should experiment with this.

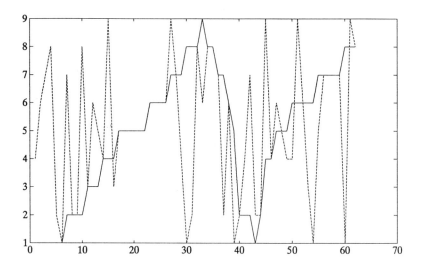

FIGURE 8. IMAGE AND DEFORMED IMAGE

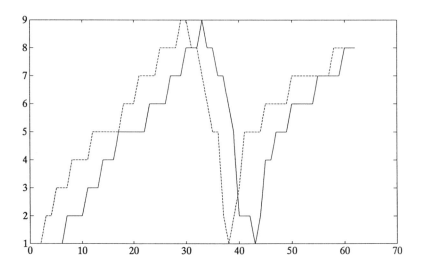

FIGURE 9. IMAGE AND RESTORED IMAGE

Remark. The random sweep strategy is probably not a good choice here. Since the M-transformations start being applied from the left it seems that it would have been better to sweep the string from left to right repeatedly. The reader may wish to try this.

Remark. The motivation behind the algorithm was DNA strings as discussed in Chapter 2. It is uncertain, however, if the method described here really applies to DNA analysis in a realistic way.

9.7. Growth patterns. So far we have only discussed static patterns; no dynamics was involved describing time development. We can, however, apply similar methods to growth patterns: one way is to interpret "simulation time" in the S.D.E.'s we have used as real time. A related approach goes as follows.

Say that we want to model the growth of *fairy rings*, the more or less annular regions formed by certain mushrooms, *Marasmius oreades*. The mushrooms do not cover the inner of a circular disk, only an annulus at its periphery. The annulus grows in size year after year.

Let us model these patterns by a lattice-based model with a 4-neighbor graph. Let us start at time $t = 1$ with a single generator of age $= 1$ in the middle of the lattice. Now at each location x, y we ask first whether it is empty conf$(x, y) = 0$. If not, we just update its age conf(x, y) by adding 1. Otherwise we count the size of its environment with non-empty sites $=$ nenv. We make conf$(x, y) = 1$ with a probability p_{nenv}, nenv $= 0, 1, 2, 3, 4$, and continue in this way.

The function growth 1 is a slow and somewhat crude program expressing this growth dynamics:

```
function age=growth1(l,totaltime,probvec)
%pattern synthesis for fairy ring growth
%l=side of square growth region
%totaltime=length of growth time interval
%probvec vector of length 5
rand('uniform');
age=zeros(l);
mid=floor(l/2);
age(mid,mid)=1;
conf=zeros(l);
conf(mid,mid)=1;
for time = 1:totaltime
        for x = 2: l-1
                for y = 2 :l-1
                        g=conf(x,y);
                        if g>0
                                age(x,y)=age(x,y)+1;
                        else nenv=(conf(x+1,y)>0)+(conf(x-1,y)>0);
                        nenv=nenv+(conf(x,y+1)>0)+(conf(x,y-1)>0);
                        conf(x,y)=rand(1)<probvec(1+nenv);
                        age(x,y)=conf(x,y);
                end
        end
    end
end
```

We display the result using seegrowth1

```
function seegrowth1(age)
l=length(age(1,:));
xv=.5.*[0 1 1 0 0];
yv=.5.*[0 0 1 1 0];
types='igrbw';
axis([1 1 1 1])
clg
hold on
for x=1:l-1
        for y=1:l-1
                typecol=1+(age(x,y)>5)+(age(x,y)>20)+(age(x,y)>35)+(age(x,y)>50)
                plot(x+xv,y+yv,types(typecol))
        end
end
```

where the different colors represent rings of different age as seen in the Plate. Note how this growth dynamics is based on ideas from Chapter 6.

A similar growth dynamics, but for star-shaped patterns, is shown in Plate 11. Modeling biological growth is a task that has attracted many scientists and has generated a huge literature. The above should be taken only for what it is: an illustration of how pattern theoretic ideas can be used in this context, and as a hint to the reader on how to extend and improve the model by more elaborate computer experiments which can lead to some exciting results.

Chapter 10

More Pattern Experiments

For many of the regular structures studied in the text the reader has been encouraged to experiment with them, not only by changing parameter values, but also by more substantial modifications. The computer experiment is a powerful tool for acquiring a real understanding of how the regular structures represent knowledge: their strengths and limitations. Below we suggest other computer experiments, some easy to do, others requiring much more effort, a whole project rather than just an example. It should be pointed out that several of them have not been carried out, as far as the author is aware, so that there may be surprises and unforeseen difficulties waiting for the experimenter.

Experiment 1

The following is an extension of the Ising model to a situation where we have more than one type of region into which the background space will be divided. Introduce generator space $G = \{1, 2, \ldots r\}$ with full information bonds $\beta_j(g) \equiv g$ on a square lattice with four neighbors, $\omega(g) = 4$. Some conditions at the boundary of the lattice will be needed; the easiest one is periodic boundaries so that we are on a discrete two-dimensional torus.

The acceptor function $A(\cdot, \cdot)$ will then be an $r \times r$ matrix with non-negative entries. Try

$$A(g', g'') = A_0(g', g'') + \delta$$

where δ is a small positive constant and

$$A_0(g', g'') = \begin{cases} a \text{ if } g' = g'' \\ b \text{ if } |g' - g''| = 1 \\ 0 \text{ else.} \end{cases}$$

Here the parameter a should be considerably larger than b.

Run stochastic relaxation on this configuration space and see if the results look reasonable. One modification of interest would be to also introduce boundary generators that tend to separate the regions, either by a single boundary generator (with its rotated versions) or different ones for separating g and $g+1$ for each g. This will require some thought on how to set up G.

Experiment 2

Synthesize an Ising model configuration. Then for each pixel (x, y) associate a random number normally distributed $N(m_0, \sigma^2)$ if $g_{xy} = 0$ and $N(m_1, \sigma^2)$ if $g_{xy} = 1$.

Here σ should be smaller than $|m_1 - m_0|$ so that the two normal distributions do not overlap very much.

If the experimenter has a device for displaying and/or printing gray-level pictures do this and study the resulting patterns for different values of the parameters a, b (in the Ising model) and m_0, m_1, σ^2.

As a substitute for gray-level displays one can use the MATLAB function *contour* although those pictures will be less informative. What sorts of textures are represented by this regular structure?

If I is the random Ising image (configuration) and $I^\mathcal{D}$ stands for the observed field of normal random variables try to use the Bayesian paradigm, perhaps using stochastic relaxation, to restore I having only observed $I^\mathcal{D}$. The ideas in Chapter 4 will be helpful.

Experiment 3

Start from an alphabet $\{a, b, c, \ldots\}$ coded into integers $\{1, 2, \ldots r\}$ and consider strings $g_1, g_2, \ldots g_n$ of length n with probabilities

$$\begin{cases} p(c) &= p_{1g_1} \cdot p_{g_1 g_2} \ldots g_{g_{n-1} g_n} \\ c &= \sigma(g_1, g_2, \ldots g_n). \end{cases}$$

The matrix $(p_{k\ell}; k, \ell = 1, 2, \ldots r)$ should be a Markov transition probability matrix so that $\sum_\ell p_{k\ell} = 1$, $\forall k$. Note that the multiplicative form of $p(\cdot)$ is as in the second structure formula (see Chapter 4) so that the synthesis and restoration method in that chapter can be applied.

Deform c into $c^\mathcal{D}$ by the deformation mechanism

$$p(g \to g') = \begin{cases} 1 - \epsilon \text{ if } g' = g \\ \frac{\epsilon}{r-1} \text{ if } g' \neq g \end{cases}$$

for each g_i. Use stochastic relaxation to restore c if $c^\mathcal{D}$ is given into a configuration c^*. Reasonable values for ϵ are in the range 10% − 30%.

To evaluate the performance of the restorations one could use the Hamming distance $d(\cdot,\cdot)$ = number of mistakes in c^* compared to c. Does the method reduce $d(c, c^\mathcal{D})$ substantially compared to $d(c, c^*)$? A reader familiar with dynamic programming can also use this technique for finding the $\hat{c}$ that maximizes the posterior probability $p(\hat{c}|c^\mathcal{D})$.

Experiment 4

The CF example in section 5.4 is really too simple. Take a fragment of some natural language, for example English, and write a CF grammar, perhaps with reflexive clauses, interrogative phrases, negating phrases, and so on.

Introduce probabilities on the rewriting rules and synthesize the language. Does it look right, or do you get too many unnatural sentences? You will find that it is not easy to get satisfactory results; this is just an indication of the fact that natural language is an extremely complicated phenomenon and that CF grammars are not realistic representations unless one makes the set of rewriting rules very large.

Experiment 5

Consider Ising model configurations on an $\ell \times \ell$ square lattice with parameters b, c. The observer loses a part of a picture $c(x, y)$, say a square E with integer coordinates $x, y : x_1 \leq x \leq x_2, y_1 \leq y \leq y_2$. This could be said to be a masking deformation mechanism; the observation will be denoted $c^\mathcal{D}(x, y)$ where $x, y \notin E$.

To do pattern extrapolation we can use stochastic relaxation where the selected sites are all in E; other values of $c(x, y)$ are not updated. This is going to simulate the conditional probability distribution of the whole picture when only $I^\mathcal{D}$ has been observed.

Synthesize c, for example with $\ell = 64, b = .1, c = 1$, and delete a square in the middle of the picture (store the whole picture c for later use). Then run pattern extrapolation as sketched above, and compare the resulting picture c^* with c, the whole pictures. How well does the algorithm perform?

Experiment 6

Try to synthesize carpet textures like the one in Figure 5 of section 3.3. One idea is to let the generators be unit vectors of orientation φ positioned at the grid points of a square $\ell \times \ell$ lattice.

A prior probability density that seems reasonable is the multivariate von Mises

$$\begin{cases} p(c) \propto exp\ b \sum_{x,y} \{cos(g_{x+1y} - g_{xy}) \quad +cos(g_{x-1y} - g_{xy}) + cos(g_{xy+1} - g_{xy}) \\ \qquad\qquad\qquad\qquad\qquad\qquad +cos(g_{xy-1} - g_{xy})\} \\ c = \text{LATTICE}(g_{xy}; x, y = 1, 2, \cdots \ell). \end{cases}$$

The exponent (equal to minus energy) has a gradient proportional to

$$sin(g_{x+1y} - g_{xy}) + sin(g_{x-1y} - g_{xy}) \\ + sin(g_{xy+1} - g_{xy}) + sin(g_{xy-1} - g_{xy})$$

so that we can synthesize it by solving the S.D.E.

$$dc(t) = dt \cdot grad + \sqrt{2}\ dw(t)$$

or discretized in time

$$c(t + dt) = c(t) + dt \cdot grad + \sqrt{2dt}w$$

where w is a square random matrix with entries i.i.d. $N(0, 1)$.

Plausible values are $b = 5, \ell = 64$ and using 500 iterations or so. The resulting configurations, displayed by drawing a unit vector with direction g_{xy} at grid point x, y, look reasonable but can one improve the representation?

Experiment 7

Consider a configuration space $\mathcal{C}(\mathcal{R}) = <G, A, \Sigma>$ where $G = \mathbb{R}$ with $\omega(g) = 4, \beta_j(g) \equiv g$ (full information bonds) and the acceptor function

$$A(\beta, \beta') = \frac{1}{1 + c(\beta - \beta')^2}$$

and $\Sigma = $ SQUARE LATTICE, say of size $\ell \times \ell$.

At the boundaries of σ let g's be fixed, for example to zero. Synthesize and display the configurations.

The energy of this A is

$$-log\ A(\beta, \beta') = log[1 + c(\beta - \beta')^2].$$

As a function of the bond difference $\delta = \beta - \beta'$ the function is convex for small δ but concave for large δ. Hence big δ-values are more likely than if the function had been convex everywhere, as is the case, for example, for a Gaussian A-function. The qualitative behavior can therefore be expected to be quite different from Gaussian random fields.

Experiment 8

The following is only intended for a reader who is interested in neural network learning and has access to software for neural nets.

Consider star-shaped images as in section 9.4. Synthesize such shapes; they will constitute the normal image ensemble. Also synthesize abnormal image ensembles. If the radii of the star-shapes are denoted $r_k, k = 1, 2, \ldots n$ apply a simple deformation mechanism by

$$r_k^{\mathcal{D}} = r_k^{\mathcal{D}}(1 + e_k); k = 1, 2, \ldots n$$

where e_k are i.i.d. normal stochastic variables $N(0, \sigma^2)$ where σ is considerably less than one.

Feed images from both ensembles into the network, telling it from which ensemble the images originate (learning with teacher), and see if the network learns to discriminate between normal/abnormal, if it can be used as an abnormality detector.

If all images are also transformed by a random rotation

$$r_k \to r_{k+\ell}, \ \ell = 1, 2, \ldots n \text{ with probabilities } 1/n,$$

and addition $k + \ell$ is understood modulo n, the learning task will be harder and the network may not learn effectively.

An extension of this would be to have more abnormal ensembles. One could be, for example, obtained by

$$r_k = \begin{cases} r_k(1 + c(k-a)(b-k)), \ a \leq k \leq b \\ r_k \text{ else} \end{cases}$$

where a and b are random and c is a positive constant (for growth deformation) or negative (for destructive deformation).

Experiment 9

The following, quite challenging, experiment in 3-D requires hardware/software for displaying three-dimensional shapes, for example as wire frames with hidden lines removed.

Introduce a template as the unit sphere in $I\!R^3$ parametrized by

$$\begin{cases} x = \cos \theta \cos \varphi \\ y = \cos \theta \sin \varphi \\ z = \sin \theta \end{cases}$$

with the longitude $\varphi \in [-\pi, \pi)$ and the latitude $\theta \in [-\pi/2, \pi/2]$. Discretize the sphere by equally spaced φ_k and (not necessarily equally spaced) θ_ℓ, so that we can speak of the points $(x_{k\ell}, y_{k\ell}, z_{k\ell})$. Avoid $\theta = \pm \pi/2$.

Let the generators consist of line segments between neighboring points, for example from $(x_{k\ell}, y_{k\ell}, z_{k\ell})$ to $(x_{k+1\ell}, y_{k+1\ell}, z_{k+1\ell})$ and to $(x_{k\ell+1}, y_{k\ell+1}, z_{k\ell+1})$. The arities will be four except the generators closest to the North and South poles for which they are equal to three.

We shall apply several similarity groups, starting with

$$S_1 = \text{translation group in } \mathbb{R}^3$$
$$S_2 = \text{scaling by } c_1, c_2, c_3 \text{ along the } x, y, z \text{ axes}$$
$$S_3 = \text{rotation group in } \mathbb{R}^3.$$

The Euler angles φ, θ, ψ can be used to parametrize the rotation matrices of S_3:

$$M(\varphi, \theta, \psi) = \begin{pmatrix} \cos\varphi \cos\varphi \cos\psi, & \sin\varphi \cos\theta \cos\psi + \cos\varphi \sin\psi, & -\sin\theta \cos\psi \\ -\cos\varphi \cos\theta \sin\psi - \sin\varphi \cos\psi, & -\sin\varphi \cos\theta \sin\psi + \cos\varphi \cos\psi, & \sin\theta \sin\psi \\ \cos\varphi \sin\theta, & \sin\varphi \sin\theta, & \cos\theta \end{pmatrix}$$

$S_4 = $ twisting around an axis which we take temporarily to be the x-axis:

$$\begin{pmatrix} x' \\ y' \\ z' \end{pmatrix} = \begin{pmatrix} 1, & 0, & 0 \\ \cos tx, & \sin tx, & 0 \\ -\sin tx, & \cos tx, & 0 \end{pmatrix} \begin{pmatrix} x \\ y \\ z \end{pmatrix}$$

$S_5 = $ bending in a plane that we take temporarily to be the (x, y)-plane:

$$\begin{cases} x' = x \\ y' = y + cx^2 \\ z' = z \end{cases}$$

For $c_{temp} = \sigma(g_{k\ell})$ deform it by letting all the 15 parameters in the group be random variables (also the x-axis in S_4 and the (x, y) plane in S_5). Display the resulting configuration

$$c = \sigma(s_5 s_4 s_3 s_2 s_1 g_{k\ell})$$

on the screen. Experiment with different parameter values and invent other similarity groups that seem natural.

This regular structure has been used to study variable amoeba shape, actually in $\mathbb{R}^4 = $ space $\times$ time, for motion as caused by a chemical gradient in a liquid. Then local deformation groups were added to the above five groups.

Experiment 10

Let the generators be smooth directed arcs in the plane, say with continuous curvature, and attribute arity two to them, $\omega(g) \equiv 2$, all of length 1. Define bond values as

$$\begin{cases} \beta_{in}(g) & = (\text{start end point } z_{in} \text{ of } g, \text{ tangent } t_{in} \text{ at } z_{in}) \\ \beta_{out}(g) & = (\text{final end point } z_{out} \text{ of } g, \text{ tangent } t_{out} \text{ at } z_{out}) \end{cases}$$

so that the bond value space B has dimension 3, location + direction.

With the connection type $\Sigma = \text{LINEAR}(n)$ and bond relation EQUAL it is natural to identify (via the identification rule R from section 4.3) a configuration $c = \text{LINEAR}(g_1, g_2, \ldots g_n)$ with the curve of length n that it represents; denote the image by $I(g_1, g_2, \ldots g_n)$.

With the similarity group $S = $ Euclidean group, consisting of translations and rotations, the minimal patterns are

$$\mathcal{P}(g_1, g_2, \ldots g_n) = \{\text{all } s \in S | sI(g_1, \ldots g_n)\} \subset \mathcal{I}.$$

Apply the following deformation mechanism $\mathcal{D}$. Divide the curve of length n by npoints equally spaced points so that we get npoints vectors $v_1, v_2, \ldots v_{npoints}$ (from point to point) of length n/npoints. Rotate each v_i by a random angle φ_i from the multivariate von Mises density proportional to

$$exp\left[b_1 \sum_{i=1}^{npoints-1} cos(\varphi_{i+1} - \varphi_i) + b_2 \sum_{i=1}^{npoints} cos\,\varphi_i\right].$$

Simulate and display the deformed curves with vertices

$$\begin{cases} z_1 = \text{starting end point of } I \\ z_2 = z_1 + s_1 v_1 \\ z_3 = z_2 + s_2 v_2 \\ \ldots \end{cases}$$

and experiment with the value of the coupling parameter b_1 and the choice of generators (arcs), perhaps straight line, circular arcs, oscillating arcs and so on.

Given a deformed curve $I^\mathcal{D}$ find the pattern $\mathcal{P}(y_1, y_2, \ldots g_n)$ of greatest likelihood. Note that there are three nuisance parameters: two for z_1 (no problem) and a rotation, fixed for all i. These nuisance parameters must be estimated.

How well does this pattern recognition algorithm perform? Display both $I(g_1, g_2, \ldots)$, assumed unknown by the algorithm, and the estimated $I(g_1^*, g_2^*, \ldots)$.

Experiment 11

The next experiment is harder and deals with image restoration in computerized tomography but in a simplified version where we have left out several technical difficulties (see below).

In a square X of size $L \times L$ define a contrast image template $I_{temp}(x, y)$ which takes large values inside some set and then rapidly decreases to zero outside the set. If the experimenter has access to a digital image like this, use it. Otherwise a synthetic image could be used, say something like

$$I_{temp}(z) = (c + \prod_i |z - z_i|^2)^{-p}; \quad z = (x, y);$$

where the z_i's are a few point in X, and p and c are positive constants.

Introduce angles $\varphi_r = \frac{\pi r}{m}; r = 0, 1, \ldots m-1$ and the m sets of lattice points, d is a small fixed constant,

$$\begin{cases} x(r, s, t) = sd \cos \varphi_r + td \sin \varphi_r \\ y(r, s, t) = -sd \sin \varphi_r + td \cos \varphi_r. \end{cases}$$

For any fixed r we let s and t range over the integers such that $(x(r, s, t), y(r, s, t)) \in X$; say $s \in S(r), t \in T(r, s)$.

The deformed image consists of the measurements

$$I^{\mathcal{D}}(r, s) = \sum_{t \in T(r,s)} I[x(r, s, t), y(r, s, t)] + e_r,$$

where

$$I(x, y) = I_{temp}[x + u(x, y), y + v(x, y)]$$

with the background deformation

$$\begin{cases} u(x, y) = \sigma \sum_{k,\ell=1}^{m} \frac{z_{k\ell}}{k^2+\ell^2} \sin \frac{\pi k x}{L} \sin \frac{\pi \ell y}{L} \\ v(x, y) = \sigma \sum_{k,\ell=1}^{m} \frac{z'_{k\ell}}{k^2+\ell^2} \sin \frac{\pi k x}{L} \sin \frac{\pi \ell y}{L} \end{cases}$$

with all $z_{k\ell}, z'_{k\ell}$ i.i.d. $N(0, 1)$. The errors e_{rs} are assumed to be independent and $N(0, \tau_{rs}^2)$. Let us choose all τ_{rs} equal, the easiest but not more realistic alternative.

The sums defining $I^{\mathcal{D}}$ are a bit like the integrals in the Radon deformation (note 4.7) and can be thought of as obtained from a Gamma camera or similar recording device.

Synthesize I and I^D, plot I^D, called the sinogram because of its wavy nature. Note that the deformed image I^D is of completely different nature from the pure image I. Now, using I^D, try to restore I.

To do this we write the posterior density over z, z' space as proportional to

$$\text{prior} \times \text{likelihood} = exp[a_1(z, z')] \times exp[a_2(z, z')]$$

The form of a_1 is simply

$$-\frac{1}{2} \sum_{k,\ell=1}^{m} [z_{k\ell}^2 + (z'_{k\ell})^2]$$

with the gradient having components like $-z_{k\ell}, -z'_{k\ell}$.

The harder part is to find the gradient of a_2. We have

$$a_2(z, z') = -\frac{1}{2\tau^2} \sum_{r,s} \{I^D(r, s) - \sum_t I_{temp}[x + u(x, y), y + v(x, y)]\}^2$$

with the shorthand notation

$$\begin{cases} x = x(r, s, t) \\ y = y(r, s, t) \end{cases}$$

and the final sum is over $r \in \{0, m-1\}, s \in S(r)$, and the second over $t \in T(r, s)$.

When we differentiate a_2 with respect to $z_{k\ell}$ we will get expressions of the form

$$\frac{1}{\tau^2} \sum_{r,s} \{\quad\} \sum_t \frac{\partial}{\partial z_{k\ell}} I_{temp}[x + u(x, y), y + v(x, y)]$$

(bracket as above in the expression for a_2) so that we need the derivatives $\frac{\partial}{\partial x} I_{temp}, \frac{\partial}{\partial y} I_{temp}$ and $\frac{\partial u}{\partial z_{k\ell}}, \frac{\partial v}{\partial z_{k\ell}}$. The latter ones are easy but the first require some approximation if $I_{temp}(x, y)$ is given in digital form, only for integral values of x, y. Do something crude, like replacing $(x + u, y + v)$ by the closest lattice point in X.

Now we can solve the S.D.E. as before. Display the I, I^D, I^*. It will take a good deal of experimentation to get satisfactory results.

Remark. We have neglected attenuation and scattering in the model; in real tomography they have to be included, which takes a good understanding of the physics of the tomography camera. A difficulty that is more important and harder to handle is the fact that our setup is intended only to describe normal variability. If we also want to include abnormal variation, say for spotting anomalies, it is necessary to extend the knowledge representation.

NOTES

Chapter 1

1.1. This refers to Galileo Galilei's 1632 treatise *Dialogo supra i due massimi sistemi del mondo* for which he was tried by the Inquisition and sentenced to imprisonment.

1.2. Laplace was made minister of finance, a position for which he was not well suited, and later fired by Napoleon: "The trouble with monsieur Laplace is that he reduces finance to the calculation of infinitesimals."

1.3. Fractals and chaos are discussed in detail in Mandelbrot (1977) and Gleick (1988).

Chapter 2

2.1. The Indian grammarian Panini, who may have lived sometime in the fifth or sixth centuries B.C., wrote what is probably the first systematic grammar in the form of many extremely concise propositions.

2.2. A reader interested in natural language patterns will find a wealth of information in Crystal (1987) presented in a highly readable style.

2.3. Markov chains are discussed in most textbooks on probability theory, for example Parzen (1960).

2.4. One of the earliest attempts to formalize line patterns was presented in Shaw (1969).

2.5. Hermeneutics is the study of interpretations and hidden meaning in the Scriptures and other texts.

2.6. A computer simulation based on a pattern theoretic model of bird flocking can be found in Heppner–Grenander (1990).

Chapter 3

3.1. Conformational analysis is the study of the geometric arrangement in space of the constituent atoms of a molecule. A popular and readable introduction to this subject is Rees (1967).

3.2. These texture patterns are from Brodatz (1966), where many others can be found.

3.3. This figure is due to S. Geman.

3.4. If the second order moments around the center (a,b) of gravity are denoted m_{20}, m_{11}, m_{02} then the eigen values of the matrix

$$\begin{pmatrix} m_{20} & m_{11} \\ m_{11} & m_{02} \end{pmatrix}; \quad m_{ij} = \int\int I(x,y)(x-a)^i(y-b)^j \, dx \, dy$$

are the lengths of the major and minor axes, and the directions of the axes correspond to the eigen vectors.

3.5. These pictures as well as the treatment of leaf shape later in the book are from Knoerr (1988).

3.6. The concept of landmarks plays a crucial role in morphometrics – the quantitative study of shape. An important reference is Bookstein (1978).

3.7. In graph theory a graph is a collection of sites (also called nodes) and segments (also called edges). Some of the sites are connected by segments while others are not. If the segments carry arrows one speaks of a directed graph.

3.8. These figures are from Geman–Manbeck–McClure (1991).

3.9. A fascinating approach to vision was given in Marr (1982), which is recommended to a reader who wants an introduction to the emerging field of "computer vision."

3.10. Stereology is a vast subject; as an introduction the reader can consult Elias (1988).

3.11. A good reference is Theocaris (1969); it is oriented towards engineering applications.

3.12. A classical and still readable presentation of elementary concepts in formal logic is Hilbert–Ackermann (1950).

3.13. Occam was a fourteenth-century English theologian who had a great influence on medieval theology. His "razor" expresses a principle of "economy of thought."

3.14. A fascinating and thought-provoking view of this contentious issue is Penrose (1989).

3.15. Turing's life and work is described in Hodges (1983), a remarkable human document, worth reading for more than one reason.

3.16. There is a vast literature on expert systems; one reference is Hunt (1986).

3.17. Colloquial (spoken) English is analyzed, for example, in Svartvik (1990).

Chapter 4

4.1. The reader needs to know only the most basic properties of groups; see for example Childs (1979). Commonly occurring groups in the text are $\mathbb{Z}$, the translation group on the integers, $\mathbb{Z}^d$ same but for a d-dimensional square lattice, $\mathbb{R}$ the translation group on the real line, $\mathbb{R}^d$ same in d-dimensional Euclidean space. Others are $E(d)$ = the Euclidean group in $\mathbb{R}^d$, $A(d)$ the affine group in $\mathbb{R}^d$. Sometimes the cyclic group on a discrete d-dimensional torus will be denoted $\mathbb{Z}_d$.

4.2. Here $\mathbb{Z}_2$ should be thought of as periodic (or a two-dimensional discrete torus) so that a site at the right boundary, for example, is a neighbor of a site at the left boundary.

4.3. The subgroup we have in mind is the center of the group S = the set of s values such that $ss' = s's$ for all $s' \in S$. If S is commutative the center is the whole group.

4.4. Recall that R is called an equivalence relation if it is 1) reflexive: cRc, 2) symmetric $cRc' \implies c'Rc$, and 3) transitive: $c_1 R c_2$ and $c_2 R c_3 \implies c_1 R c_3$.

4.5. The concepts from statistical mechanics that are mentioned in the text also appear in the study of Markov random fields; see for example Kinderman and Snell (1980).

4.6. A homeomorphic mapping $h : X \leftrightarrow X$ is one that is one-one and continuous in both directions. It preserves the topology of X.

4.7. The mapping expressed by the line integral is called a *Radon transform*; it plays a major role in many medical scanning methods.

4.8. The quotient space X/S, of a set X with respect to a similarity group S has as elements sets of mutually equivalent x-values modulo S.

Chapter 6

6.1. Analog computing may be having a renaissance; see the thought-provoking book by Mead (1988).

6.2. There is a large literature on the Ising model; one reference is McCoy–Wu (1973).

6.3. Early instances of this methodology go back to the 1950s when physicists developed techniques of this type. More recently they have been applied to spatial statistics and image processing; see Besag (1974, 1986) and Ripley (1988, 1991), where many more references can be found.

6.4. For Markov chains see e.g. Feller (1957).

6.5. See e.g. Feller (1957), p. 114.

6.6. These figures were obtained by P. Schay.

6.7. Simulated annealing in this context was first used in Geman–Geman (1984).

6.8. ICM is due to J. Besag.

6.9. To see this diagonalize $M = O^T DO = O^T D^{1/2} OO^T D^{1/2} O = N^2$ with $N = O^T D^{1/2} O$, where $D^{1/2}$ is the diagonal matrix whose entries are the square root of the entries in D.

6.10. The leaf shapes are provided by A. Knoerr.

6.11. This series is known as the Karhunen–Loève expansion.

6.12. A detailed description of the inference algorithm can be found in Grenander–Chow–Keenan (1991) together with lots of restorations. Some, but not all of them, can be regarded as successes. The ones that failed happened because the deformation model was not sufficient in those cases.

6.13. This fact is well known from the theory of Markov processes. An intuitive justification for it can be found in Grenander (1993), pp. 137–140. It should be noted, however, that conditions are needed to make the statement valid; this is too complicated to be discussed here. A Wiener process $W(t)$ taking values in some $I\!R^d$ is a stochastic process where increments $W(t + h) - W(t)$ are $N(0, hI)$, $I = d \times d$ identity matrix, and where increments over disjoint intervals are stochastically independent.

6.14. This is joint work with K. Manbeck. Pictures illustrating the results can be found in Grenander–Manbeck (1993).

6.15. This is described in Grenander–Miller (1994), where proofs of convergence for the jump-diffusion process are given.

Chapter 7

7.1. APL = A Programming Language (developed by K. Iversen) is an unusual language due to K. Iversen with its own notation, differing essentially from most others. Many versions exist, but they have almost the same syntax.

7.2. MATLAB is developed by Math Works, Inc., and is available for many platforms. The code described below was developed on a 386 PC, and a diskette with some of the programs is available; information about it can be obtained by writing to the author.

7.3. MATHEMATICA was developed by Wolfram Research, Inc.

7.4. The printer used for this purpose was a Hewlett-Packard Ink Jet; it is easy to use and fairly inexpensive.

Chapter 8

8.1. This code was written by P. Schay.

8.2. The general linear group GL(2) consists of all non-singular 2 × 2 matrices with the group operation understood as matrix multiplication.

8.3. MAP estimation = $\underline{M}$aximum $\underline{A}$ $\underline{P}$osteriori estimation; just maximize the posterior density.

Chapter 9

9.1. These programs are from P. Schay.

9.2. This is from Amit–Grenander–Piccioni (1991).

9.3. A Bernoulli process x_t, $t = 1, 2, \ldots, x_n$, is simply an i.i.d. sample from a Bernoulli variable x, $P(x = 1) = p$, $P(x = 0) = 1 - p$.

REFERENCES

This list represents only a small fragment of the vast literature on pattern research and mentions only references referred to in the text.

Amit, Y., U. Grenander and M. Piccioni (1991), *Structural image restoration through deformable templates*, J. Am. Stat. Assoc., 376–387.

Anson, B. J. (1963), *An Atlas of Human Anatomy*, Saunders, Philadelphia.

Besag, J. (1974), *Spatial interaction and the statistical analysis of lattice systems*, J. Royal Stat. Soc. **B, 36**, 192–236.

Besag, J. (1986), *On the statistical analysis of dirty pictures*, J. Royal Stat. Soc. **B, 48**, 259–302.

Bookstein, F. L. (1978), *The Measurement of Biological Shape and Shape Change*, Lecture Notes in Biomathematics, Springer-Verlag, vol. 24, New York.

Bouthan, J. (1968), *ABC of the EKG*, Philips Tech. Library, Eindhoven, The Netherlands.

Brodatz, P. (1966), *TEXTURES: A Photographic Album for Artists and Designers*, Dover, New York.

Childs, L. (1979), *A Concrete Introduction to Higher Algebra*, Springer-Verlag, New York.

Crystal, D. (1987), *The Cambridge Encyclopedia of Language*, Cambridge University Press, Cambridge.

Elias, H. (1983), *A Guide to Practical Stereology*, Karger, Basel, Switzerland.

Feller, W. (1957), *An Introduction to Probability Theory and its Applications*, Second ed., John Wiley & Sons, New York.

Geman, D. and S. Geman (1984), *Stochastic relaxation, Gibbs distributions and the Bayesian restoration of images*, IEEE Trans. Pattern Analysis and Machine Intell. **6**, 721–741.

Geman, S., K. M. Manbeck and D. E. McClure (1991), *A comprehensive statistical model for Single Photon Emission Tomography*, in Markov Random Fields: Theory and Applications, eds. R. Challeppa and A. Jain, Academic Press, New York.

Gleick, J. (1988), *Chaos: Making a New Science*, Penguin Books, New York.

Grenander, U., Y. Chow and K. M. Keenan (1991), *HANDS: A Pattern Theoretic Study of Biological Shapes*, Springer-Verlag, New York.

Grenander, U. and K. Manbeck (1993), *A stochastic shape and color model for defect detection in potatoes*, Amer. Stat. Assoc. **2, No. 2**, 131–151.

Grenander, U. and M. I. Miller (1994), *Representations of knowledge in complex systems*, J. Royal Stat. Soc. **56**, 549–603.

Grenander, U. (1993), *General Pattern Theory*, Oxford University Press, Oxford.

Heppner, F. and U. Grenander (1990), *A stochastic linear model for coordinated bird flocks*, in The Ubiquity of Chaos, ed. S. Krasner, Am. Assoc. Adv. Science, Washington, D.C..

Hilbert, D. and W. Ackermann (1950, original German ed. 1928), *Principles of Mathematical Logic*, Chelsea, New York.

Hodges, H. (1983), *Alan Turing: The Enigma*, Simon and Schuster, New York.

Hunt, V. D. (1986), *Artificial Intelligence and Expert Systems Source Book*, Chapman and Hall, New York.

Kinderman, R. and J. L. Snell (1980), *Markov Random Fields and their Applications*, American Mathematical Society, Providence.

Knoerr, A. (1988), *Global models of natural boundaries*, Rept. in Pattern Analysis, Div. Appl. Math., Brown University, Providence.

Mandelbrot, B. (1977), *The Fractal Geometry of Nature*, Freeman, New York.

Marr, D. (1982), *Vision: A Computational Investigation into the Human Representation and Processing of Visual Information*, Freeman, San Francisco.

McCoy, B. and T. Wu (1973), *The Two-Dimensional Ising Model*, Harvard University Press, Cambridge.

Mead, C. (1988), *Analog VLSI and Neural Systems*, Addison-Wesley, Reading.

Mosteller, F. and D. L. Wallace (1964), *Inference and Disputed Authorship: The Federalist*, Addison-Wesley, Reading.

Owen, J. (1972, original ed. 1856), *The Grammar of Ornaments*, Van Nostrand, New York.

Parzen, E. (1960), *Modern Probability Theory and its Applications*, John Wiley & Sons, New York.

Penrose, R. (1989), *The Emperor's New Mind: Concerning Computers, Minds, and the Laws of Physics*, Oxford University Press, Oxford.

Propp, V. (1975, original ed. 1928), *Morphology of the Folk Tale*, University of Texas Press, Austin.

Rees, D. A. (1967), *The Shape of Molecules*, Oliver & Boyd, Edinburgh.

Ripley, B. D. (1988), *Statistical Inference for Spatial Processes*, Cambridge University Press, Cambridge.

Ripley, B. D. (1991), *The use of spatial models as image priors*, in Spatial Statistics and Imaging, ed. A. Possolo, Inst. Math. Stat, Hayward.

Shannon, C. (1949), *The Mathematical Theory of Communication*, University of Illinois Press, Urbana.

Shaw, A. C. (1969), *A form description scheme as a basis for picture processing systems*, Inf. and Control **14**, 9–52.

Svartvik, J. ed. (1990), *The London–Lund Corpus of Spoken English*, Lund University Press, Lund.

Theocaris, P. S. (1969), *Moiré Fringes in Strain Analysis*, Pergamon Press, London.

Thompson, D'Arcy W. (1961, original ed. 1917), *On Growth and Form*, Cambridge University Press, Cambridge.

Weyl, H. (1952), *Symmetry*, Princeton University Press, Princeton.

A reader who wishes to delve deeper into the subjects discussed in this book is recommended to consult Ripley (1988) for a readable account of spatial processes and A. Possolo, ed. (1991) Spatial Statistics and Imaging, Inst. Math. Stat., Hayward, California. Another recent source is J. Appl. Statistics, Vol. 16, No. 2, 1989, edited by K.V. Mardia. A comprehensive but technically difficult presentation of pattern theory is Grenander (1993).

INDEX

abnormality detection, 100, 133
acceptor function, 95
affinity, 95
APL, 146
arity, 82

behavior patterns, 30
bird flocking, 31
bistable configuration, 194
bond, 82; coordinate, 83; external, 84; internal, 84; structure, 82; structure group, 82; value relation, 85
boundary: image, 93; pattern, 119

C, 147
C. elegans, 65–66
closed patterns, 34
cloth, 46
clustering pattern, 45
computational modules, 75
configuration, 83
configuration space, 85
connector, 84
context free language, 20, 107–8
contrast patterns, 29
controlled randomness, 38
coordinates, local, 42
correspondence, 56
coupling connector, 87
crystal lattice, 43
crystals, 42
curve: image, 108; patterns, 70

deformations, 25; automorphic, 97; background, 98; contrast, 97; image, 97; incomplete, 97; indirect, 98. *See also* deformed template

deformed template, 91; probabilistically, 91
doctrines, 32

electro-physiological recordings (EKG, EEG, EMG), 26–28
energy, 95
equilibrium distribution, 115

Federalist Papers, 15
finite state: automaton, 160; language, 106; machine, 17
FORTRAN, 147
frozen image, 96
fundamental cell, 44
fur, 46

Galilean simplification, 6
generator, 81; coordinate, 83; diagram, 82; index, 82
grammatical phrase, 106
growth pattern, 199

Hamming distance, 25
hand pattern, 54
heteromorphic deformation, 97
homeomorphic map, 127
homeomorphism, 58
homomorphism, 90
hypothesis formation, 69

ice/water pattern, 48
identification rule, 91
image, 91; cold, 96; extrapolation, 99; hot, 96; restoration, 99; segmentation, 99; understanding, 100
invariance, 81
Ising model, 112

jittering, 90

knowledge representations, 5

landmarks, 55, 132
language, 13, 106. *See also* context free language
leaf shapes, 51, 125
Levenshtein distance, 26
linear configuration, 101
line patterns, 29
Lissajou pattern, 152
literary patterns, and the Federalist Papers, 15
logic, uncompromising, 129

Markov random field, 97
MATHEMATICA, 146
MATLAB, 146
metric pattern, 96
minimal pattern, 93
mitochondria, 68
Moiré fringes, 71
molecular chains, 23
monotonic, 86
motion pattern, 21
multimodal inference, 190
multiple objects, 140

open patterns, 33
order/disorder, 4
organelles, 67
ornaments, 8

Pap smear, 66–67
parallel: computer, 148; logic, 130
partition function, 95
pattern, 93; algebra, 81; analysis, 99; dynamics, 26; formalism, 77; inference, 99; interference, 70; recognition, 99; synthesis, 98; understanding, 76
plant patterns, 28
platonic solids, 64
probabilities, 94
Ptolemaic system, 22

regime, 28
regularity: global, 85; local, 85; relaxed, 96; rigid, 97; of structure, 91
rewriting rule, 107

sassafras leaves, 52
sentence, 106
set patterns, 29
shapes, 48
shift-operation, 89
similarity group, 81
simulated annealing, 119
sinogram, 62
snake shapes, 168
speculation patterns, 71
splines, 64
star-shape, 187
stochastic relaxation, 115
stomachs: curvatura major, 48; curvatura minor, 48; variations in form, 49
structure formula: first, 85; second, 95
symmetry, 81

temperature, 95
template, 58, 72. *See also* deformed template
textures, 44
theories, as "atoms of thought," 73–74
tomography, 61
track patterns, 29
transformed images, 57
trimer configuration, 130

understanding: extrinsic, 100; intrinsic, 100; pattern, 76

weave patterns, 38, 175; basket, 39; herringbone, 40; satin, 41; twill, 40
wood grain, 46

X-rays, 56, 58, 134

yarns, 23

Library of Congress Cataloging-in-Publication Data

Grenander, Ulf.
   Elements of pattern theory / Ulf Grenander.
      p.   cm. — (Johns Hopkins series in the mathematical sciences)
   Includes bibliographical references (p.   –   ) and index.
   ISBN 0-8018-5187-4 (hc : alk. paper). — ISBN 0-8018-5188-2 (pbk. : alk. paper)
   1. Pattern perception. 2. Pattern recognition systems. I. Title. II. Series.
Q327.G727   1996
511.3'3—dc20   95-15912
                                                                                                                                  CIP